Flowers

Evolution of the Floral Architecture of Angiosperms

Flowers

Evolution of the Floral Architecture of Angiosperms

Guillaume Tcherkez
Université Paris-Sud
France

Preface by

Pierre-Henri Gouyon
Professor, University of Paris-Sud
Institut National Agronomique, Ecole Polytechnique
Director, Ecologie, Systematique et Evolution Laboratory of CNRS-UPS at Orsay

Science Publishers, Inc.

Enfield (NH), USA   Plymouth, UK

CIP data will be provided on request.

SCIENCE PUBLISHERS, INC.
Post Office Box 699
Enfield, New Hampshire 03748
United States of America

Internet site: *http://www.scipub.net*

sales@scipub.net (marketing department)
editor@scipub.net (editorial department)
info@scipub.net (for all other enquiries)

ISBN 1-57808-311-7

Published by arrangement with Dunod, Paris.

Ouvrage publié avec le concours du Ministère français chargé de la Culture - Centre national du livre
Published with the support of the French Ministry of Culture

Translation of: ***Les Fleurs,*** Évolution de l'architecture florale des Angiospermes, Dunod, Paris, 2002.
French edition: © Dunod, Paris, 2002.
ISBN 2 10 005844 4

Published by Science Publishers, Inc. Enfield, NH, USA
Printed in India.

Preface

The flower is undoubtedly the most remarkable and most central characteristic of the Spermatophytes. The study of the flower is therefore central to botany. Botany is defined primarily as the science of plants. In this respect, it encompasses everything that concerns the study of plants, from the molecule to the biosphere. However, the term has taken on a meaning that is often slightly antiquated, especially in countries with a Latin-based language. The botanist is considered to be a dreamer whose favourite activity is to wander in a green environment, exclaiming at rare specimens, bizarre flowers that he or she presses between sheets of paper. The herbarium, the microscope, and the Latin system of naming are the attributes of such a botanist. Still, when we observe a researcher extracting DNA from a sample (even a green sample) in a test tube, or one programming arcane equations on a computer, let us remember that they too may be botanists.

Botany has taken on all these forms throughout its history. The systematic knowledge of the diversity of species, their classification, and their nomenclature remains of course an essential part of this science. But an understanding of the intimate functioning of the plant, particularly of the flower, whether it is on the biochemical, cellular, molecular, or organism scale, must also be considered a part of botany. Moreover, to understand the form and function of the flower, root, leaf, or stem, we must use concepts not only from physiology and developmental biology, but also from ecology and evolutionary sciences, integrating genetic, demographic, and biogeographical aspects in a process that includes experimental study as well as observation and mathematical and/or statistical modelling. Clearly, botany is a "complete" science. Following various approaches, it no longer simply supposes an encyclopaedic knowledge of form but demands an understanding of mechanisms. It has, in sum, undergone the formidable evolution of all the life sciences during the 19th and 20th centuries, and it has been an engrossing and important component of that evolution.

Genetics was discovered by Mendel on peas, then rediscovered by de Vries, Tschermak, and Correns on other plants. Linear regression, a classic tool of statistics, came from another study on peas, conducted by Galton. The concept of mutation developed from the study of *Oenothera* by de Vries and maternal inheritance from the study of *Silene* flowers by Correns (we now know that these are characters determined by the mitochondrial genome). The concepts of genotype and phenotype came from Johannsen's study of beans. The onion provided images of chromosomes during cell division and it was on maize that Barbara McClintock discovered transposons. Botany has made a fantastic contribution to modern biology.

Throughout this evolution of scientific knowledge, we can see that the fields of major discoveries have shrunk in size. The biologist's approach has become increasingly reductionist. Today, it has reached an extreme at which we can practically study biology without knowing anything about what happens at a scale beyond that of the cell, or even the DNA molecule and what surrounds it. It is unlikely, and certainly undesirable, that this trend will continue. Knowledge of the genome reveals little about the organism and still less about the ecosystem. After the necessary dive into the microscopic, biology must re-emerge into its integrative processes. Botany will necessarily be among the fields that allow such changes in scale. We hope that this work, which shows how the discipline of botany today allows us to understand the constitution of flowers, will help the reader to perceive the interest of the whole.

Here it is a question of science, botany, and flowers. The flower is the sexual organ of plants. Linnaeus focused on this erotico-pedagogical organ, understanding that, being the reproductive organ, it was undoubtedly the most essential element of the individual. Darwin articulated his reflections on the study of flowers in *The Different Forms of Flowers on Plants of the Same Species* (1877) and in his treatises on orchids. The understanding of the evolution and development of flowers has revealed surprising phenomena about the way in which genes determine the structure of flowers and about the ecological and evolutionary phenomena that are at work. The flower is undoubtedly the key to the realm of plants. Let us enter and marvel at it.

Pierre-Henri Gouyon

Professor, University of Paris-Sud, Institut National Agronomique, Ecole Polytechnique

Director, Ecologie, Systematique et Evolution Laboratory of CNRS-UPS at Orsay

Contents

Introduction

From ancient times, human societies have paid particular attention to flowers, on at least two levels, the aesthetic and the symbolic. For example, the Egyptians and the Far Eastern civilizations often represented lotus flowers for ornamental purposes. The lotus was also a symbol of purity, because the flowers bloomed just above the surface of the water and were not sullied by mud, which was considered to represent vice and misery. The builders of architectural wonders were also inspired by floral motifs. Such motifs are found in Arabian palaces, such as that found in the garden city of Alhambra, in Grenada. In fact, three types of complementary motifs—floral, geometric, and graphic—were abundantly used under the arcades, along door frames, and even on some walls. This architectural complementarity links flowers, geometry, and poesy, suggesting that the flower is not considered an isolated entity but seen through its architecture and its meaning, the whole composing a harmonious motif that touches the sensitivity of the observer. The flower thus acquires the dimension of a signifier.

The signified became scientific only from the Renaissance onwards, when there emerged a passion to label and classify everything. These were the first steps towards systematics. In the decades that followed, more classifications were developed, benefiting from the formalization of Linnaeus and the reflections of Jussieu, as well as the botanical descriptions of Redouté, who lent his hand to the artistic aspect of plant systematics, as did artists such as Desportes through their detailed work in their still life paintings.

How do we define a flower? Two classic definitions are recognized, based on the arrangement or nature of the floral organs.

The first definition: A set of sterile and fertile reproductive parts arranged generally in whorls, the whole being axillated by a bract.

The second definition: A set of reproductive organs comprising one or more ovules, which are defined by their integuments.

Each definition has its advantages and limitations. The first emphasizes the arrangement in **whorls**, which is, however, not the rule (Magnoliaceae,

Ranunculaceae), and the existence of the **bract**. There remains some uncertainty about the nature of the floral parts, and we do not see any inherent reason to exclude the reproductive parts of certain Pteridospermates[1] or "seed ferns", in which, by convention, the status of bract can be given to the leaf immediately below the reproductive structures. The second definition also does not exclude the structures of Pteridospermales from "flowers", since at least some of them have ovules very similar to those of Angiosperms.

In the Gymnosperms (from *gymnos*, naked, and *spermum*, seed), there are clearly flowers comparable to those of Angiosperms, except, by definition, in the "nakedness" of ovules. For example, the female cones of pine (*Pinus* sp., Pinaceae) are clearly inflorescences: each scale is a flower, comprising only two ovules and a bract. This bract forms a protruding tongue in Douglas (*Pseudostuga* sp.), where it is thus clearly visible.

The definition of flower that we adopt in this work is based on the "typical" angiosperm architecture, arranged in whorls or in spirals, with sepals, petals, or tepals, stamens and carpels, all of which are present in the axil of a bract.[2]

Angiosperms actually appeared in the middle of the Cretaceous period, around 130 million years ago, in the form of dicotyledonous plants (Le Thomas,1999). The search for the first genera of Angiosperms is an ongoing one. Some fossils tend to call into question the date of appearance of this group even though, dating from the Triassic or the Jurassic, they are only hypothetically associated with the Angiosperms. The uncertainty concerning such fossils could be due to the fact that we cannot verify even the characteristic of Angiosperms: angiospermy or "hermetic" sealing of carpels. However, a fossil found recently in northeastern China, of the genus *Archaefructus*, presents a branch carrying follicles containing seeds (Sun et al., 1998), pushing the origin of Angiosperms back to the Upper Jurassic.

Angiosperms today account for 220,000 species, as opposed to 750 counted for the Gymnosperms (Bawa, 1995). The oldest fossils of monocotyledons date from the Turonian, about 90 million years ago (Gandolfo et al., 1998). At the end of the Cretaceous, most of the groups already existed. The origin of Angiosperms remains a subject of debate: the Gnetales appear very similar to the Angiosperms from an anatomical point of view, but phylogenetic analyses of genes *rbcS, rbcL* (coding respectively the small and large subunits of Rubisco), *cox1* (cytochrome-oxidase), and *atpA* (ATP-synthase) seem

---

[1] The Pteridospermales, called *seed ferns*—even though the term *seed* is not suitable because, although prothalli are seen, a sporophytic embryo is never seen—constitute an entirely fossil group, whose representatives look like ferns and produce structures comparable to ovules. The Pteridospermales, a paraphyletic group, do not belong to the Spermatophytes, precisely because of the absence of an "authentic" seed.

[2] The bract is sometimes difficult to observe, as for example in certain varieties of tulips (*Tulipa* sp., Liliaceae), because the attachment zone is very low. However, this aspect must be taken into consideration in the floral architecture, as we will explain later on.

Diagram of a fossil branch of follicles of *Archaefructus*. The bar represents around 5 mm (modified from Sun et al., 2000).

to show that, apart from the fact that the Gymnosperms are monophyletic, the nearest clade of the Gnetales would be Conifers (Bowe et al., 2000; Chaw et al., 2000).

The fundamental difference between the flowers of Gymnosperms and those of Angiosperms lies by definition in the covering of ovules: never complete in the former, and "absolute" in the latter. There are, however, some exceptions: first, after fertilization, the scales of pines tighten as they dry, hermetically enclosing the ovules; second, some Angiosperms present incompletely closed

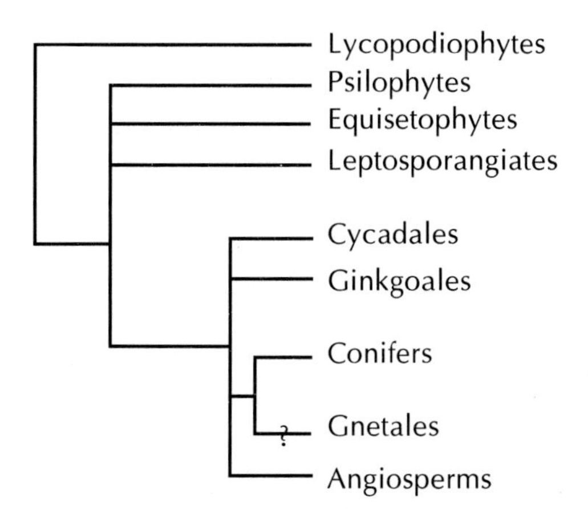

Simplified phylogenetic tree showing some major plant groups. The question mark indicates uncertainty about the place of Gnetales (see text), which were first placed close to Angiosperms and today are found to be close to Conifers (as presented in this figure).

carpels (e.g., Amborellaceae, Resedaceae, which remains disputed). The case of *Amborella*, a genus that stands alone at the base of phylogenetic trees (see Chapter 3), suggests that carpels remaining open at maturity and even after seed formation represent the ancestral state (Raven et al., 2001).

Moreover, the flowers or inflorescences of Angiosperms show high degrees of specialization, which is not the case with Gymnosperms in general. In the latter, in fact, the pollen, which frequently has balloons, is dispersed by the wind. In the Angiosperms, we observe on the contrary many strategies of pollen dispersal and collection. The floral structures reflect the **fertilization strategies** adopted by different species. These aspects will constitute a central theme in the pages that follow. That is why we begin with basic aspects of the floral architecture and then tackle the evolution of the design of floral organization: (1) from "ancient" groups to "modern" groups of Angiosperms and (2) in relation with the diversification of fertilization strategies within groups and the biological problems that these strategies imply.

1

Inflorescence Architecture

In this chapter, the aspects relating to groups of flowers or **inflorescences** are briefly developed and only the basic concepts of description of these structures are given, in order to address evolutionary issues later.

1.1. FLOWERS AND INFLORESCENCES

1.1.1. Location of Inflorescences on the Plant

Two kinds of Angiosperms are discussed here—ligneous and herbaceous.

a) Ligneous plants

The position of inflorescences depends greatly on the **architecture** of the organism. The architecture of woody plants can be formalized by an architectural model (Halle and Oldeman, 1970). There are essentially 21 architectural models, which are distinguished from one another by the nature of axes observed (single, differentiated, non-differentiated, or mixed), the modes of growth (basi- or acrotonous, epi- or hypotonic, rhythmic or continuous), the modes of branching (bifurcation, monopodial or sympodial growth), and the position of flowers (epitony, hypotony, and meristems at work). To clarify these ideas, we take three examples:

— Agave (*Agave americana*, Agavaceae). The agave forms a woody, non-ramified axis (single axis) in which the apical meristem forms the inflorescence. There is thus an end to the development since the axes can no longer grow. The plant dies after some time. The resulting architecture responds to the Holttum model (Fig. 1.1, part 1). In this case, the inflorescence is apical and of an indefinite type.

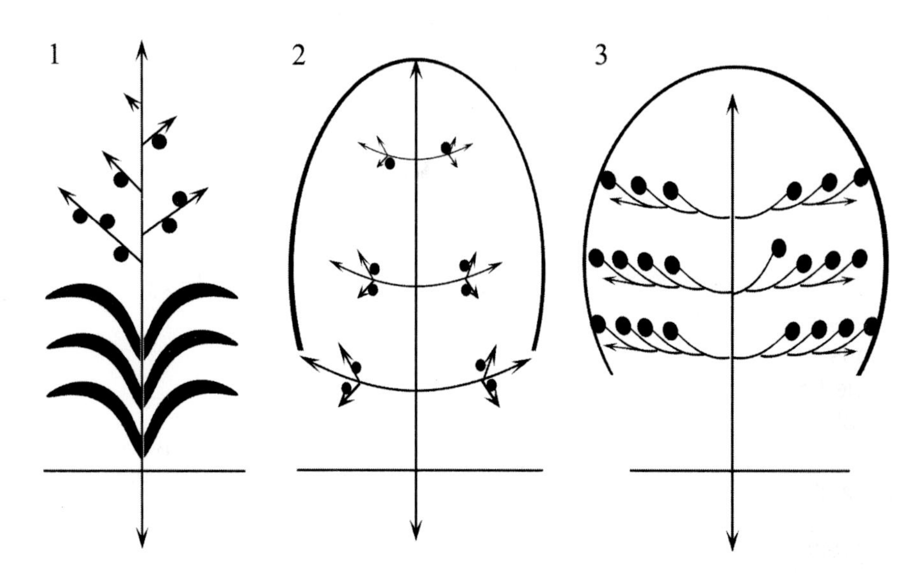

Fig. 1.1. Three examples of architectural models of woody plants: (1) Holttum model, (2), Rauh model, and (3) Petit model. The flowers are indicated by a black dot and the apical meristems by arrowheads.

— The sycamore (*Acer pseudoplatanus*, Sapindaceae earlier Aceraceae) presents a rhythmic growth (based on the seasons), monopodial, orthotropic, and ramified (the secondary axes, borne on the primary axes, are opposite, as are the tertiary axes, borne on the secondary axes). The flowers are borne on the tertiary axes (or quaternary axes, if any). There is not always a preferential orientation for flower formation, i.e., there is no evident floral epitony or hypotony. In terms of nomenclature this architecture is described by the Rauh model, and it is undoubtedly the most common on earth (Fig. 1.1, part 2).

— Magnolia (*Magnolia* sp., Magnoliaceae) develops lateral branches by juxtaposition of leafy axes with definite growth and terminal flowering. This process is called "growth by substitution" (Koriba, 1958; Halle et al., 1978). The architecture corresponding to magnolia is associated with the Petit model (Fig. 1.1, part 3).

We will not further pursue the description of architectural models because if there is a lesson to be drawn from this study, it is that inflorescence position does not follow a general rule. Inflorescences are borne on axes I, II, III, or IV, they may be epitonic or hypotonic. However, the position is important because it has repercussions, for example, on accessibility and visibility to pollinators (or more generally the pollen vector) or extent of investment in reproduction (number of flowers). For example, the agave is destined to die after flowering, so we see a gigantic inflorescence. The plant undoubtedly uses all its potential for growth and morphogenesis in its ultimate realization, reproduction. In terms of trade-off between vegetative development and reproduction, evolutionary strategies differ from one species to another. The perpetuation of the species is based on short-lived individuals having an unbalanced trade-off leaning towards reproduction (agave) or long-lived individuals having a balanced trade-off.

b) Herbaceous plants

There is also a wide variety of positions occupied by inflorescences in herbaceous plants: terminal (e.g., tulip) or non-terminal (e.g., black nightshade), carried by axes I (tulip) or axes of a higher order (black nightshade), epitonic or hypotonic, etc.

1.1.2. Origin of Floral Meristem

Inflorescences are groups of flowers and constitute units of the first order in the description of flowering. The triggering of inflorescence formation corresponds to the transition from the vegetative to the reproductive state of an apical axial meristem: this is "floral induction". The meristem is thus an inflorescence meristem. It will form the inflorescence and the floral meristems that give rise to flowers. The distinction between inflorescence meristem and floral meristem is not arbitrary: in *Arabidopsis thaliana* (Brassicaceae), the mutant *cal* forms inflorescences lacking flowers. There seems to be a close genetic control of vegetative-inflorescence transitions and inflorescence-floral transitions, so that two types of genes can be distinguished, i.e., one for inflorescence identity, the other for floral identity. For example, mutants such as *cal* have an indefinite cluster even though there is no flower formation. The absence of expression of genes of the second type continues the inflorescence state.

Besides, these genetic transitions depend on the photoperiod, the age of the plant, and other factors that will not be addressed here.

1.2. ARCHITECTURE OF INFLORESCENCES

1.2.1. From the Solitary Flower to the Inflorescence

Some plants do not have inflorescences and directly form a flower from the apical axial meristem. This is the case (generally) with tulip (*Tulipa* sp., Liliaceae). There are many gradations between the solitary flower and inflorescences with a large number of flowers (Asteraceae, for example). A single species may have a varying number of flowers within an inflorescence, and that

Genes and Meristems

Generally, there are two classes of genes involved in the early control of meristems: those corresponding to "common characters" and those corresponding to the differences (orientation towards a floral destiny) between the three types of meristem (i.e., inflorescence, floral, and axial).

The establishment and maintenance of the apical axial meristem is controlled by *pnh* (*pinhead*) and *zll* (*zwille*), the activities of which are strictly embryonic, and *stm* (*shoot meristemless*) and *wus* (*wuschel*), expressed continuously. Mutations on one of these four genes lead to a sometimes highly significant reduction of the apical axial meristem, while mutation on a gene of the series *clv* (*clavata*) leads to an enlargement of the meristem. In particular, *clv* and *stm* seem antagonistic. The double mutant *clv stm* has apparently normal meristems at the beginning of development but this state, which is subsequently not stable, leads to hypertrophied or very small meristems. There are interactions between the other genes, some of which remain to be detailed, for example, the mutation *wus* is epistasic on *clv*, suggesting that *clv* regulates *wus* negatively or even that *wus* is necessary to the triggering of *clv* action (Weigel et al., 1996). In any case, *stm* and *clv* seem to be two very important genes for the maintenance of the apical axial meristem.

The transition of a meristem towards the floral state is controlled by the genes *lfy* (leafy) and *ap-1* (*apetala-1*). For example, transgenic plants of *Arabidopsis thaliana* expressing constitutively *lfy* or *ap-1* (under promoter 35S) form normal flowers in places where they habitually develop branches. Inversely, the mutants *lfy* or *ap-1* (*floricaula* and *squamosa* in *Antirrhinum majus*) are incapable of forming flowers and develop branches with indefinite growth in place of flowers. The genes upstream of the "transition" genes determine in particular the identity of floral organs (see Chapter 3).

Other genes determine other aspects of development: some mutants, such as *terminal flower* in *Arabidopsis*, have a definite cluster terminated by a flower. The function of this gene in the wild type seems thus to be to inhibit transformation of the inflorescence into a flower. The control of the gene in the plant, however, remains unclear. In any case, it is possible that this type of gene is not expressed in plants with defined inflorescences (of the cyme type) but that hypothesis is still to be confirmed (Vallade, 1999). Moreover, genes of inflorescence identity, not yet studied in detail, modify the expression of other genes (e.g., *knat-1* in *A. majus*) such as those controlling phyllotaxis. Indeed, the vegetative-inflorescence transition is accompanied by a change in organ arrangement in *Antirrhinum*, decussate for the leaves and spiral for the flowers.

depends greatly on the environmental conditions or the time of inflorescence development.

An inflorescence does not signify equivalence between its constituent flowers: in some cases a structural or temporal differentiation of the flowers occurs.

a) Structural differentiation

A typical example is that of Asteraceae. In arnica (*Arnica montana*), two types of flowers are seen on a single inflorescence: those that are central and yellow and those that are peripheral and orange. The latter form a tongue, which is why they are called **ligulate**, while the former are called **tubular**. They do not have the same internal structure: the tubular flowers have stamens, the ligulate flowers have an asymmetrical corolla (three petals more developed than the others and forming the ligule) and may lack stamens. This strategy of grouping-differentiation of flowers somewhat imitates a solitary flower of standard organization, and the pollinators may take them for such a flower.

b) Temporal differentiation

In the common teasel (*Dipsacus fullonum*, Dipsacaceae), the blooming time is not the same from one stage of the inflorescence to another, so that the inflorescence seems to flower in "sequence". The biological significance of this process remains uncertain. However, various hypotheses can be formulated. The non-simultaneous flowering of all the flowers increases the probability that each flower blooming at a given moment will be visited by a pollinator. Moreover, the flowering is extended over a longer period of time, which also increases the probability of visit, in the case of competition between blooming plants.

The number of flowers per inflorescence is thus not arbitrary. It probably results from two selection pressures with opposite effects: an increase in the number of flowers allows it to form more seeds after fertilization and helps attract pollinators, but it increases the probability that the pollinator will successively visit two flowers from the same inflorescence. From a genetic perspective, if the particular species is supposed to be self-compatible, this behaviour will lead to fertilization between flowers of a single plant (or **geitonogamy**) and favour inbreeding depression. It is thus likely that the number of flowers in an inflorescence is decided by a compromise between these two aspects (see below).

1.2.2. Diversity

Various types of inflorescences exist in the Angiosperms, and they are named according to the determinate or indeterminate nature of their growth. When the apical meristem of the inflorescence produces flowers without interruption (until the inflorescence growth is terminated), it is called an **indeterminate** or **racemose** inflorescence. In the contrary case, the functioning (or flower formation) of the inflorescence meristem ceases in favour of the meristem or meristems flanking it, and this is called a **determinate** inflorescence.

Variations are found within each type:

a) Indeterminate inflorescence

In a **raceme**, the inflorescence axis forms flowers with identifiable pedicels (e.g.,Wisteria, Fabaceae). It is called a **spike** when the pedicels are imperceptible (e.g., lavender, Lamiaceae). Sometimes the pedicels of the first flowers formed by the inflorescence meristem are longer than those of the last flowers formed, so that all the flowers end up on the same plane. These are called a **corymb** (e.g., in Achilles, Asteraceae, the compound inflorescences are arranged in a corymb). When the inflorescence axis develops lateral axes, themselves organized into racems, it is called a **panicle** (e.g., lilac, Oleaceae; Fig. 1.2).

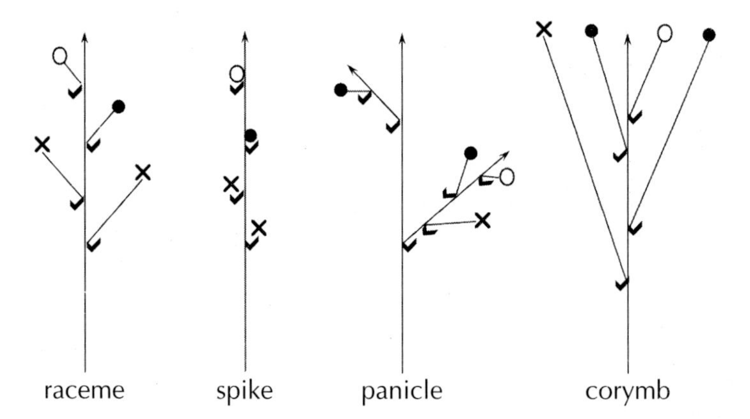

Fig. 1.2. Indeterminate inflorescences. Solid circle, open flower. Open circle, flower bud. X, faded flower. Angle, bract.

b) Determinate inflorescence

In the case of a determiante inflorescence, the term *cyme* is often used. If the two meristems (there are generally two) flank the apical meristem, it is called a **biparous** cyme (e.g., begonia, Begoniaceae). If only one meristem develops the inflorescence, it is called a **uniparous** cyme.[3] When the uniparous cyme is

[3] Biparous and uniparous cymes are the equivalent of unichasial and dichasial vegetative ramifications respectively.

very contracted (e.g., comfrey, Boraginaceae), it is called "scorpioid" because of its resemblance to a scorpion's tail (Fig. 1.3).

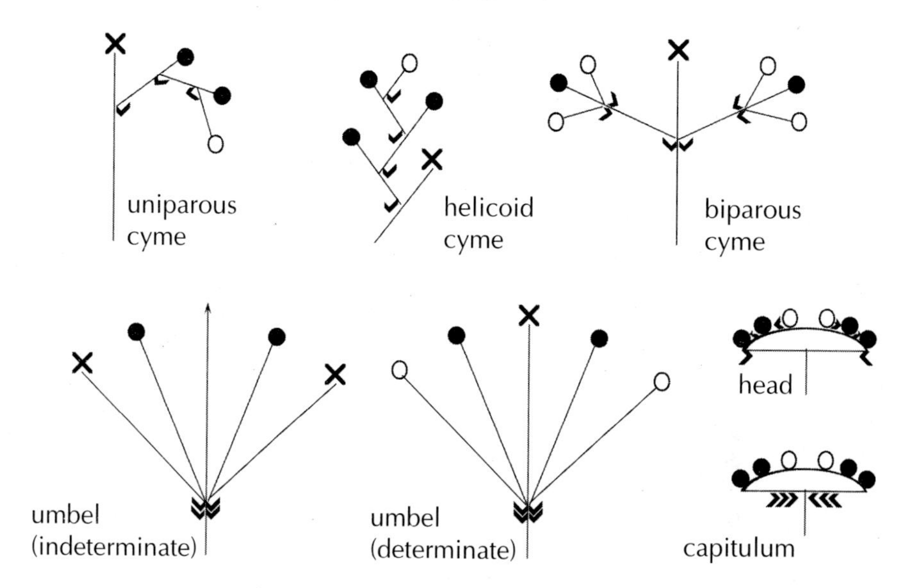

Fig. 1.3. Cymes, umbels, and compact inflorescences. Solid circle, open flower. Open circle, flower bud. X, faded flower. Angle, bract.

c) More complex inflorescences

Umbels

An **umbel** is an inflorescence in which all the floral pedicels (or secondary inflorescences) are attached at a single point. The secondary inflorescences may also have an umbel structure, in which case they are called **umbellules**. The umbel and umbellule structure is very common in the Apiaceae (earlier Umbelliferae). From a developmental point of view, umbels form in two ways: the central flower of the umbel is either the last formed or the first formed. In the first case, the umbel is a kind of raceme in which the floral pedicels are attached all at the same point (there is thus no internodal growth). In the second case, it is a cyme without internodal growth.

Heads and capitula

The mode of formation of very contracted inflorescences cannot be described after the development of the inflorescences. At least two types of condensed groups of flowers are found. In the **head** type, each flower has its bract adjacent to it (e.g., common teasel, Dipsacaceae): the inflorescence is only a very dense juxtaposition of small flowers. The inflorescence axis is often dilated, which gives the whole structure a globular appearance. In the **capitulum** type, the flowers are located so close to each other that the bracts are all pushed to the base or periphery of the inflorescence (e.g., marguerite, Asteraceae). These inflorescences are similar to the indeterminate type.

Spadices
Some spikes are partly enveloped by a leafy structure, often highly modified (pre-leaf, see below) and thus the inflorescence axis is dilated, as for example in the Arum (Araceae). Such an inflorescence is called a **spadix**.

Compound inflorescences
Most particularly seen is the "cluster of cymes", i.e., **thyrsus** (e.g., grapevine, Vitaceae), and a corymb of capitula (e.g., Achilles, Asteraceae).

In some cases it is difficult to identify a structure as either a cyme or a raceme, as especially in the gladiolus (Iridaceae). The inflorescence of this plant is in fact a cyme, the ramification of which alternates from left to right, and thus the architecture takes on a zigzag appearance. This cyme, called "helicoid", strongly resembles a spike.

The **formalization** of inflorescences in diagrams uses a combination of symbols: a vertical line for the inflorescence axis, a small angle for a bract, a solid circle for a flower bud, an open circle for an open flower, and X for a faded flower (see Figs. 1.2 and 1.3). The distinction of different stages of floral evolution by means of various symbols is necessary to distinguish the cymes from racemes.

1.2.3. Pre-leaf

The development of the inflorescence meristem begins with, among other things, the establishment of pre-leaves along the axis of a branch, after the development of the axillary leaf. There are generally two pre-leaves in the Tricolporates (see Chapter 3) and more rarely a single pre-leaf. In the Aristolochiales, there is only a single pre-leaf, as in Monocotyledons (Vallade, 1999).

1.3. EXAMPLES OF EVOLUTION OF INFLORESCENCE ARCHITECTURE

1.3.1. Reduction and Modification

A typical example consists of the determinate inflorescences of families such as Fagaceae or Betulaceae (Fig. 1.4), which are reduced and modified into **cymules**. The lines that allow us to discover the trace of the original cymes are the bracts and pre-leaves. For example, in hornbeam (Betulaceae), we find the pre-leaf, bracts of order 1, and bracts of order 2. However, only flowers of

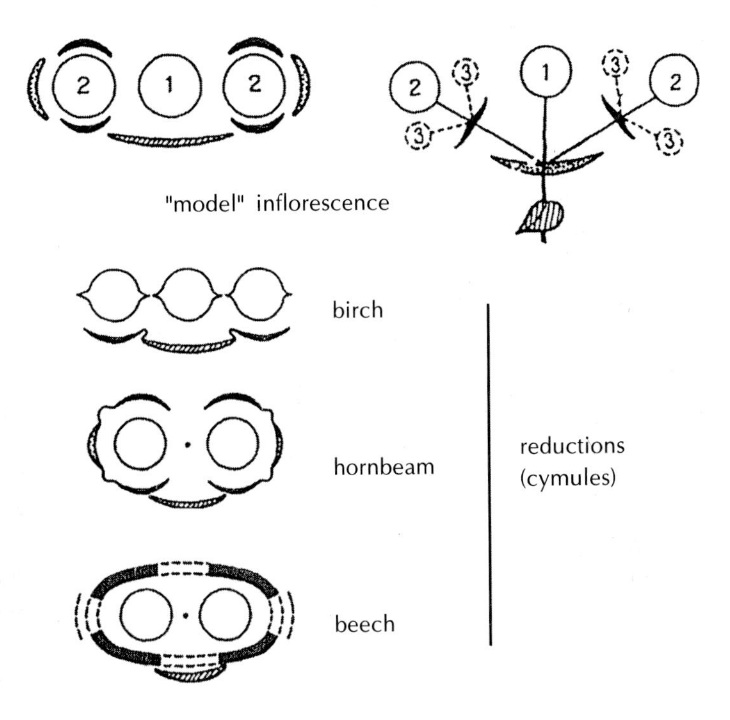

Fig. 1.4. Examples of cymules of Fagales (birch, hornbeam, beech). 1, 2, 3: flowers of orders 1, 2, and 3.

order 2 exist. In birch, the bracts of order 2 have disappeared. On the other hand, the three flowers are present. This extreme inflorescence reduction aggregates the cymules with reduced flowers into **catkins** (pendant or erect), a character that is also found in the Fagales.

Inflorescences of Euphorbiaceae are remarkable for their boxed structure, in the form of **false flowers**, associated with a floral reduction. In fact, an element of the inflorescence of *Euphorbia* sp. comprises male flowers nearly reduced to stamens, surrounding a female flower with protuberant carpels, at the end of a **gynophore**. The stamens (i.e., male flowers) are in fact arranged in uniparous cymules (with bracts reduced to the state of filamentous bracteoles), so that the stamens further inside are more mature (and thus larger). Petals are most often absent.

Finally, it has been supposed that the flower of Cruciferae (Brassicaceae) is the result of extreme reduction of an inflorescence of the cyme type. However, as we will see in the following chapter, the genetic data do not really confirm this hypothesis.

1.3.2. Aggregation and Differentiation

As has been indicated, the **capitulum** inflorescences of Asteraceae, referred to as rayed, are examples of aggregation linked to **differentiation** into ligulate and tubular flowers (see above).

There are, however, cases of aggregated inflorescences without differentiation, as in the Globulariaceae or the Chloranthaceae. This last family is also interesting in that:

— the inflorescences are in dense spikes,
— the flowers have no perianth (except in *Hedyosmum*), and
— the stamens (from 1 to 5) and the single carpel (the flowers are either hermaphrodite or not) are closely associated with a bract that supports them.

This structure is thus not very different from that of **cones** of Conifers, and the term **convergence** is used (Fig. 1.5). In fact, the Piperales (Chloranthaceae and Piperaceae) have overall an inflorescence architecture that is **derivative** of a **dense spike**, which is most compact in Chloranthaceae.

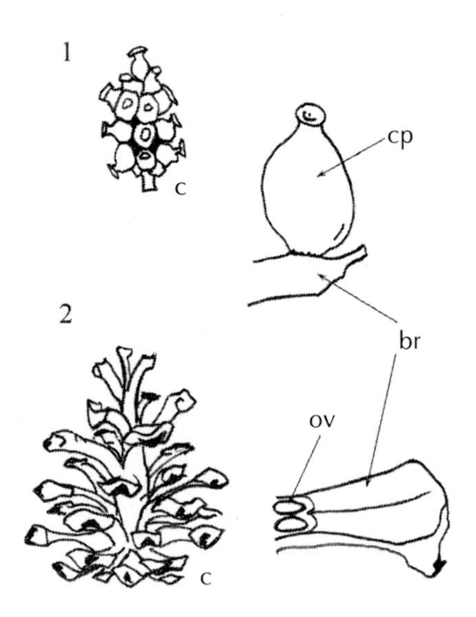

Fig. 1.5. Structural convergence of female inflorescences (cones, c) of Pinaceae (2) and of Chloranthaceae with unisexual flowers (1). In both cases, the flower is very small—one carpel (cp) in the Chloranthaceae or two ovules (ov) in the Pinaceae—and has a bract (br) that supports it.

1.3.3. Theoretical Approach (Asclepiadaceae)

In this section, we discuss theoretical aspects relating to the inflorescence architecture (i.e., elaboration of models). The theory concerning the architecture of individual flowers is treated in Chapter 2. The study of inflorescence architecture or design is based partly on problems of resource allocation and male and female reproductive success (or fitness) (denoted respectively by f_m and f_f). The first hypothesis of models is that the inflorescences have a hierarchical structure: the flowers are arranged in structures of order 1 (umbellules, for example) denoted by U_1, which are in turn grouped in units of order 2 (umbel of umbellules) denoted by U_2. The male reproductive success is generally difficult to measure (see Chapter 5), except in some plants such as the Asclepiadaceae, where it is easy to quantify the flows of pollen, grouped into pollinia, and where a direct relationship has been shown between the rate of pollinia removal and male reproductive success (Broyles et al., 1990). This is why *Asclepias* are frequently used as a study model.

What are the parameters that regulate inflorescence size? We have already said (see section 1.2.1) that inflorescence size is likely to be the result of selection via the male function, because an increase in the number of flowers increases the inflorescence attractiveness and the export of pollen. The hypothesis that this selection developed mostly by means of the male function is known as the Bateman principle. However, the increase in the number of flowers increases the probability of pollen reception. This indicates that it is not always certain that selection via female function is negligible; likewise, regarding the problem of available resources for reproduction it should be understood that the size of the inflorescence also depends on a trade-off between vegetative and reproductive growths. Specifically, it has been shown in the Orchidaceae and the Gentianaceae that male and female reproductive success is linked to inflorescence size. Moreover, there is some ambiguity between one study and another about the meaning of "inflorescence size" because it may refer to the size of units U_1 or U_2.

In the model that follows, which concerns *Asclepias*, the term "size" is attributed to U_1, because the number of flowers per umbellule seems effectively fixed genetically, which is not so for the number of umbellules or the total number of flowers on the plant (Fishbein et al., 1996).

The total fitness of a plant is expressed as the sum of male and female reproductive successes. If z is the number of flowers per umbellule, μ the number of umbellules, and $f_m(z)$ (or $f_f(z)$) the male (or female) reproductive success of an individual flower, then the total reproductive success rate is expressed as follows:

$$s(z) = f_f(z)\, z\mu + f_m(z)\, z\mu$$

A variant having size z' and a number of umbellules μ' in a population of N resident individuals having the parameters z and μ has a total reproductive success expressed as follows:

$$s_z(z') = f_f(z')\, z'\mu' + Nf_f(z)\, z\mu\, f_m(z')\, z'\mu' / [f_m(z')\, z'\mu' + (N-1)\, f_m(z)\, z\mu]$$

The first term pertains to the total female reproductive success, and the second pertains to the total male reproductive success. The success of a pollen grain of a male comprises among other things the development of the seeds that it generates, which is why the expression $Nf_f(z)\, z\mu$ is used. The seeds produced are, moreover, proportionate to the relative quantity of pollen grains of the variant in the total population, which is:

$$f_m(z')\, z'\mu' / [f_m(z')\, z'\mu' + (N-1)\, f_m(z)\, z\mu]$$

If N is sufficiently large, then:

$$s_z(z') \sim f_f(z')\, z'\mu' + Nf_f(z)\, z\mu\, f_m(z')\, z'\mu' / N f_m(z)\, z\mu$$

If, moreover, we assume that the quantity of resources available for reproduction is limited to R and is distributed between the number of umbellules and the number of flowers per umbellule, according to $R = \mu z$, we have:

$$s_z(z') \sim f_f(z')\, R' + f_f(z)\, f_m(z')\, R' / f_m(z)$$

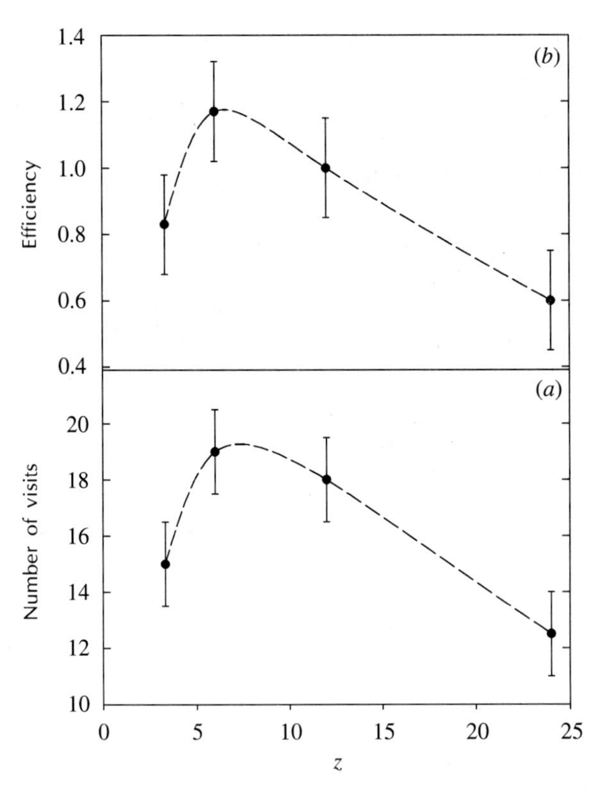

Fig. 1.6. Number of visits (*a*) and efficiency of pollinium dispersal (*b*) as a function of umbellule size *z* in *Asclepias tuberosa* (modified from Fishbein et al., 1996).

An **evolutionarily stable strategy** or ESS (Maynard Smith, 1982) is one that, once established, cannot be "encroached on" by a variant, i.e., it has a fitness or reproductive success that is greater than or equal to other possible strategies. A necessary condition for a life trait (here the size z), in z^*, to be an ESS, is that:

$$[\partial s_z (z')/\partial z']_{z = z' = z^*} = 0 \tag{1}$$

In this case, Equation 1 is equivalent to:

$$-[\partial \ln f_f(z')/\partial z']_{z' = z^*} = [\partial \ln f_m(z')/\partial z']_{z' = z^*} \tag{2}$$

where ln is the natural logarithm. Thus, the size z^* is an ESS if the relative variations of male and female reproductive success are opposed. But this condition, although it seems to apply generally, is in fact linked to the particular model that has been chosen, which is used to calculate the fitness $s_z(z')$. (In contrast, the process that, leading us to Equation 2, uses an expression of the fitness of a variant of trait z' in a population of individuals carrying the trait z is commonly used in adaptive dynamics.)

In any case, Equation 2 can be used to predict z^*, which is not generally the maximum size (possible from a biological point of view), but an **intermediate umbellule size**, as has been confirmed by studies in *Asclepias* from tracing the $f(z)$ curves (Fishbein et al., 1996; Fig. 1.6). Other results derived from models supposing that the cost c of production of an umbellule is fixed and that $s_z(z')$ is frequency-independent (the preceding model supposed a frequency dependence) show that z^* could be high, low, or intermediate, depending on the values of c and the $f(z)$ functions (Schoen and Dubuc, 1990).

2

Floral Architecture and Morphogenesis

The structure of flowers is sometimes quite complex. That is why we begin this chapter by studying the structure of a relatively simple flower that constitutes a model for molecular biology: *Arabidopsis thaliana* (Brassicaceae). Recent developments in this discipline have made it possible to understand partly the determination and evolution of the floral structure through the control of genetic expression and phylogeny.

2.1. FLORAL ARCHITECTURE

2.1.1. Description of *A. thaliana*

Arabidopsis thaliana, which belongs to the family Brassicaceae, has a typical floral structure called "type 4". The flower is found at the widened end of a **peduncle** (sometimes incorrectly called pedicel), called the **receptacle**. The flower is arranged in 4 whorls (Fig. 2.1). The outermost, which is the **calyx**, is made up of **sepals**. Then there are the **petals**, which are white, which together constitute the **corolla**. Further inside, we find the **stamens**, which together form the **androecium**, and the **carpels** (here 2 united carpels) forming the **gynoecium**. The stamens are composed of two parts: basally the elongated **filament** and apically the **anther**, which contains pollen grains. The external and internal extensions of the filament at the level of the anther are respectively called the **groove** and the **connective**. The gynoecium is composed of a swollen part or **ovary** and a slightly dilated extremity or **stigma**. In quantitative terms, there are 4 sepals, 4 petals, 6 stamens, and 2 carpels (united). We also see in some cases the presence of nectaries (depending on the taxa).

Since the architecture of the flower is in successively enclosed cycles (calyx, corolla, androecium, gynoecium), i.e., in whorls, the structure is known

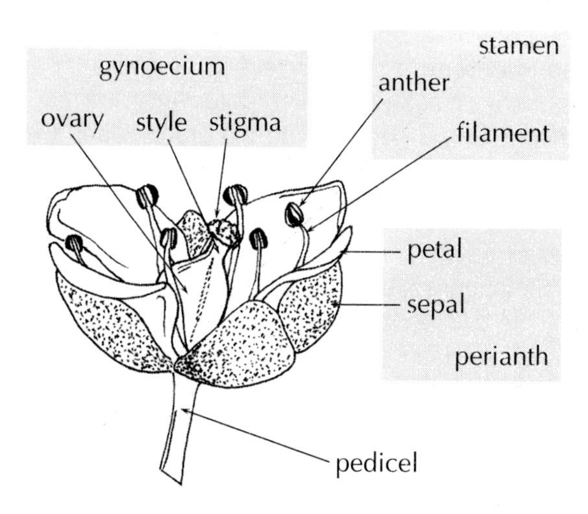

Fig. 2.1: Simple diagram of the flower of *Arabidopsis thaliana*.

as **cyclic**. It will be seen that this is not always the case and there are especially flowers with a **spiral** construction (acyclic). In that case, it is a **tetracyclic** flower because it is of type 4.

When the flower is observed from a distance, it looks as if it has axial symmetry (the axis would be perpendicular to the plane of the flower and collinear with the axis of the peduncle) and the flower is thus called **actinomorphic**. But this is only what appears from a distance. Because of the number and particular arrangement of the stamens, this flower has only a single plane of symmetry (it is thus **zygomorphic**) or two perpendicular planes of symmetry (**bisymmetrical** flower).

2.1.2. Generalization

a) Perianth

By definition, the perianth (from *peri*, around, and *anthos*, male) is made up of the calyx, which comprises the sepals, and the corolla, which comprises the petals. The sepals, ordinarily greenish and hard, protect the flower when it is closed. The petals, often soft and coloured, serve to attract pollinators if there are any. However, the attribution of these functions is not always exact, especially when the sepals have a petaloid appearance, the tepals (term designating sepals as well as petals) being implicated simultaneously in protection and the attraction of pollinators. In this case, some use the term **perigonium**. However, this term is not suitable since it remains less explicit than perianth.

Remark: The parts with a protective role, particularly the sepals, may include an **indumentum,** a set of sclerified or glandular trichomes (e.g., *Solanum sisymbrifolium,* Solanaceae). Moreover, there may be one or more protective **thorns** borne generally on the bracts, as in the common burdock (*Arctium minus,* Asteraceae), where the tip of the thorn is curved into a hook.

The pigments of petals are found in the epidermis or in the "mesophyll" (internal cells). The colour resulting from the presence of pigments is sometimes modified by the existence of intercellular spaces of the mesophyll or by epidermal papillae. The latter greatly modify the reflection of light rays and thus affect the colour (Chapter 4). These papillae may also be involved in the diffusion of floral fragrances. The non-petaloid sepals have a histological structure similar to that of leaves, with palisade and spongy parenchyma. Finally, the veins of sepals are often attenuate (narrow towards the extremity), while those of petals are bifid or ramified.

Lobes of the perianth could be united (connate). When the sepals are united the flower is called **gamosepalous** (e.g., Primulaceae) and when the petals are united the flower is called **gamopetalous** (e.g., Campanulaceae). Inversely, the dialysepalous and dialypetalous flowers have free sepals and petals respectively. The overall arrangement and position of sepals and petals in the flower bud constitutes **aestivation** (Fig. 2.2). As the flower begins to develop, the sepals, with their imbricate arrangement (e.g., Dipterocarpaceae, Clusiaceae, Theaceae) or valvular arrangement (e.g., Mimosaceae, Rhizophoraceae), protect the other floral organs, which are formed subsequently. Sometimes, it is the first sepal formed that protects the whole flower by enveloping it partly (e.g., Caesalpiniaceae, Gesneriaceae). Generally, the petals develop late and sometimes quite suddenly just before the opening of

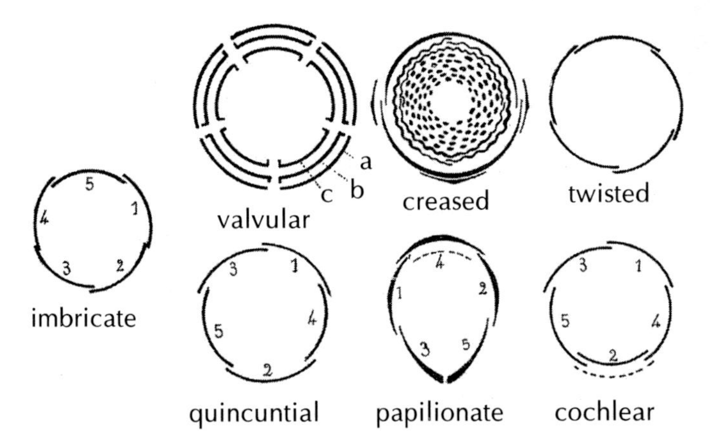

Fig. 2.2: Different types of aestivation (modified from Bach et al., 1963).

the flower. The abundance of intercellular spaces (see above) is a consequence of this rapidity of development. The petals thus take on an irregular arrangement. In some cases, they have a protective function and do not have this "typical" delayed development. This happens when short sepals protect the flower only in the early stages of floral development. The petals thus have an imbricate aestivation (Myrtaceae) or valvular aestivation (Araliaceae, Mimosaceae, or Vitaceae).

After anthesis, the perianth lobes enter into autolysis (the case of Convolvulaceae has been particularly studied, see Chapter 4) or persist: this is the case mostly with sepals (e.g., Lamiaceae), while the petals degenerate. For example, after fertilization, the sepals form two wings on the fruit in the Dipterocarpaceae (from the Greek *di*, two, *pteron*, wing, and *karpos*, fruit). Finally, the sepals may sometimes be ephemeral and fall very quickly, even before the flower opens (Papaveraceae).

Remark: In some species, there is a supplementary whorl of floral parts that look like sepals: the **epicalyx** (e.g., Malvaceae). To a lesser degree these parts protect the flower at the beginning of its development. Finally, the whorl of sepals may double: e.g., in the Cistaceae, some species with 5 sepals have a double whorl of 3 + 2 sepals.

In some cases, the parts of the perianth have secretory functions.

— Protective secretions of the sepals: the young floral organs are covered by a mucilage produced by the glandular trichomes carried by the sepals (Verbenaceae, Solanaceae, Gesneriaceae).
— Production of nectar: many genera have nectariferous glands at the base of the petals (Fig. 2.3). Nectariferous glands carried on the inner surface of sepals are more rare (Malvales).
— Hyperstigma: the internal tepals of the female flowers of Monimiaceae have a secretory activity and constitute the region in which pollen grains are deposited (Fig. 2.4). These begin to germinate *in situ*, and the pollen tubes then are directed towards the carpels. The internal tepals in these species constitute a **hyperstigma**.

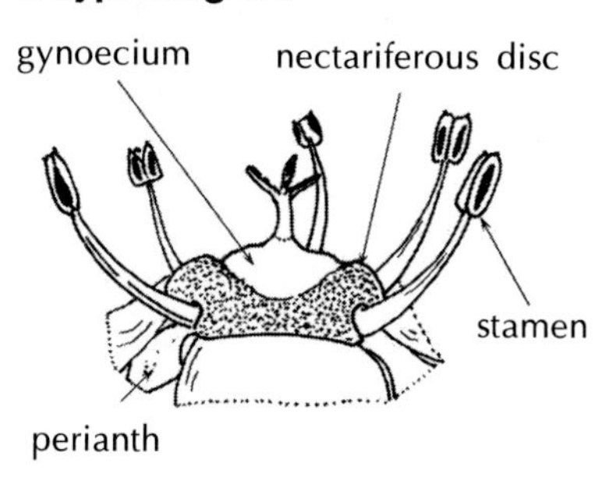

Fig. 2.3: Diagram of a flower of Celastraceae showing the nectariferous disc.

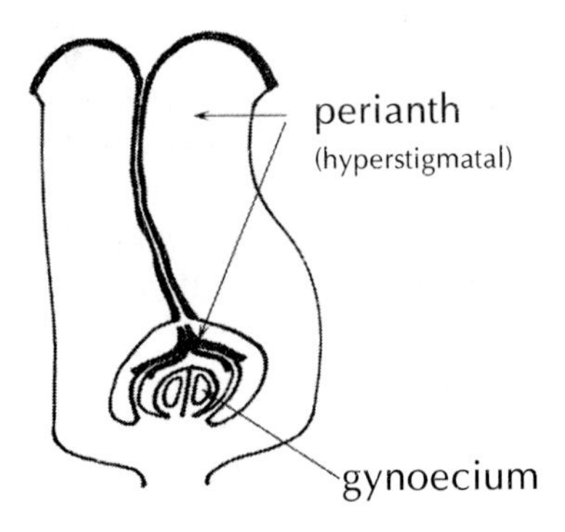

Fig. 2.4: Diagram of a flower with a hyperstigma in *Hennecartia* (Monimiaceae) (modified from Endress, 1994)

b) Androecium

Definitions

The androecium by definition constitutes all the parts with a male function (**stamens**) and those that are derived from them (**staminodes**, etc.). The stamen is made up of the filament and the anther, which contains the pollen grains. The anther includes prolongation of the filament: the connective on the inside and the groove on the outside (Fig. 2.5). The filament may be short, the anthers then being hidden within the flower, or long, so that the anther protrudes from the flower. In general, the anthers are only about a centimetre long, no doubt for reasons of mechanical stability. The *Strelitzia* or bird of paradise (Strelitziaceae) represents an extreme case, with anthers of 5 cm length. The

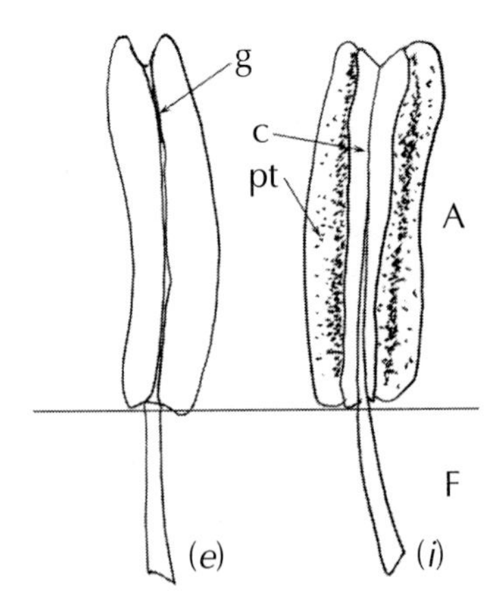

Fig. 2.5: Diagram of a stamen of lily. View of ventral or internal side (*i*) and dorsal or external side (*e*). A, anther; F, filament; g, groove; c, connective; pt, pollen theca.

filament has a cribro-vascular bundle, which generally extends to the connective. This bundle is formed simultaneously with (in the usual case) or after prolongation of the growing filament (Poaceae). In the case of Poaceae, the filament shows an extraordinarily rapid growth—up to half a centimetre in a few minutes; this organ holds the record for fast growth among the plants. The development of stamens is a case of exemplary allometry (increased allometry to the benefit of the filament).

Each anther is made of 4 **pollen sacs** joined in pairs, on either side of the elongated part of the filament (i.e., the connective). At maturity, the walls of the pollen sacs break longitudinally so that the two adjacent pollen sacs form only a single cavity, the **pollen theca**. Thus, a mature stamen has two pollen thecae (Fig. 2.6).

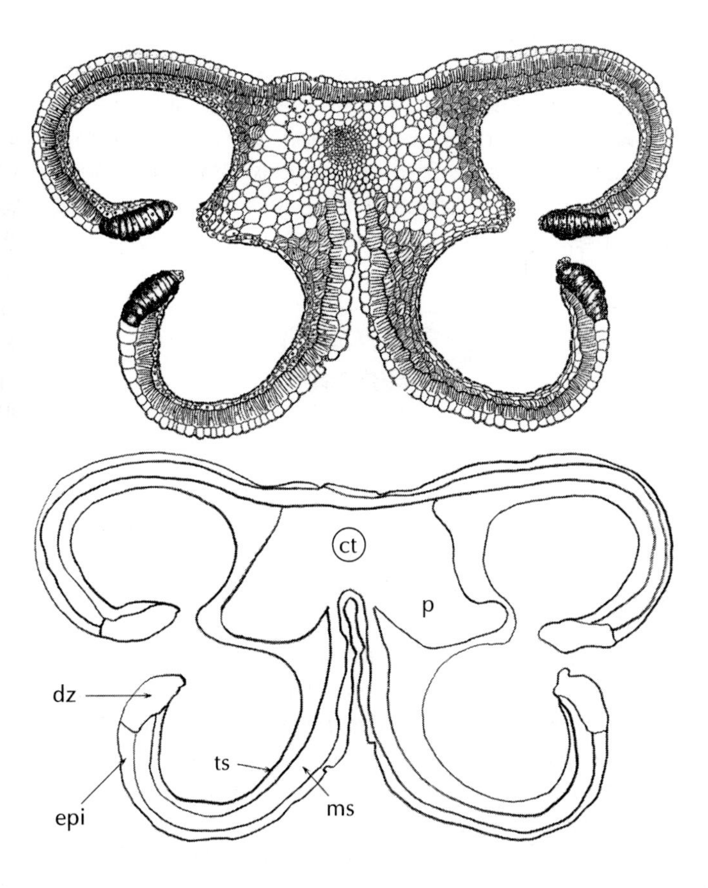

Fig. 2.6: Illustration of a cross-section of a lily stamen (top) and corresponding diagram for interpretation (bottom). epi, epidermis; ms, mechanical stratum; ts, transitory stratum; dz, dehiscence zone; p, parenchyma; ct, conducting tissue. The stamen is directed in such a way that its dorsal side is towards the top (modified from Jones et al., 1950).

Arrangement

The pollen sacs (and their groove of longitudinal dehiscence) can face outward, in which case the stamens are called **extrorse** (e.g., Ranunculaceae), or inward, in which case the stamens are called **introrse**. These configurations have consequences for the connective and the groove: when the stamens are extrorse, the groove is thin, and the connective eventually widens; this is why the stamens are called **ventrifixed**. Inversely, the introrse stamens are usually **dorsifixed**. However, this correspondence is not absolute: in tropical plants of the genus *Montinia* (Escalloniaceae), the stamens are extrorse and dorsifixed (Endress, 1994). In any case, let us note that the stamens, like all floral organs, are **dorsiventral**.

The stamens may be joined together (connate) by the filaments or by the anthers, in which case the flowers are called **syngenecious or synantherous** (e.g., Asteraceae). When the stamens are joined by the filaments, they may form a virtual **staminal tube** (e.g., Malvaceae), or even a **staminal cylinder** when the flower considered is unisexual. The joining of stamens may go as far as the formation of a sort of giant stamen, the **synandrium**, with a single circular dehiscence cleft (e.g., *Stephania japonica*, Menispermaceae) (Fig. 2.7). Finally, the stamens may be adnate to the petals (e.g., Solanaceae) and sometimes to the carpels (e.g., Aristolochiaceae).

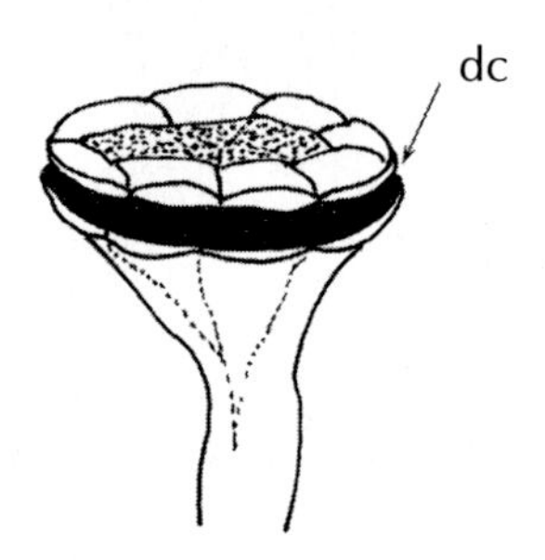

Fig. 2.7: Diagram of a synandrium (*Stephania*). dc, dehiscence cleft.

In flowers with whorled arrangement of parts, the stamens are organized in one or several whorls and are found opposite to the sepals (frequently, in which case stamens are called **antisepalous**) and/or to the petals (rare, in which case they are **antipetalous**, e.g., Primulaceae).

Sometimes the stamens are extremely modified to such an extent that they support only two masses of aggregated pollen grains, the **pollinia**. This is typically the case with Orchidaceae and Asclepiadaceae (see Chapter 4). Finally, it should be pointed out that although it is sometimes easy to decide the nature of floral organs, and in particular regarding the difference between petals and stamens, there are **false organs**; for example, in the Caryophyllaceae, the organs usually called "petals" are in fact profoundly modified stamens. The marked morphological difference between petals and sepals is a derived characteristic (see Chapter 3).

Dehiscence

There are several types of dehiscence of anthers (Fig. 2.8). As has already been suggested, frequently a longitudinal dehiscence cleft is seen at each lobe. The two clefts of the lobes may be united at the tip of the stamen. More rarely, each pollen sac has a cleft (*Strelitzia*). Besides these, there are **valvular** dehiscences (holes having a valve) or **poral** dehiscences (terminal pores, e.g., Ericaceae). Some anthers have pollen sacs divided into saclets. The saclets open towards the exterior by common or independent dehiscence clefts. An extreme case is the presence of many independent saclets, each opening by means of a pore. This type of stamen has lost the bipartite anther/filament organization. Such an organization is found especially in the Rafflesiaceae.

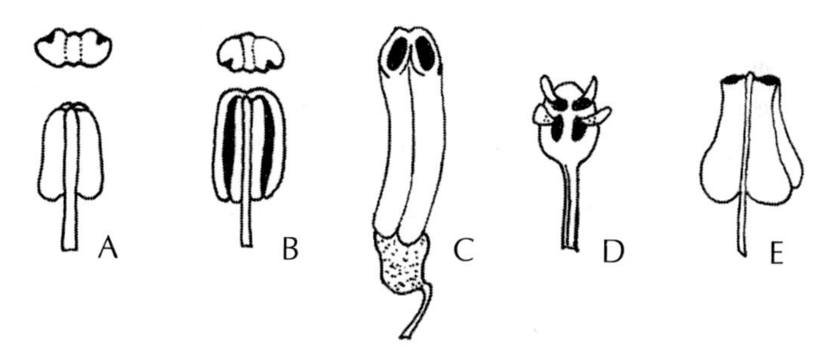

Fig. 2.8: Different types of staminal dehiscence: A, extrorse; B, introrse; C (*Daniella*) and E (*Rhododendron*), poricidal; D (*Cinnamonum*), valvular.

Other stamen characters

The stamens sometimes have particular characters linked to the realization of a precise function (Fig. 2.9):

— Trichomes of filament: the trichomes are believed to have several reproductive roles: to guide pollinators, to constrain the position of the pollinator during its progression into the flower, to obstruct uninvited insects (which steal nectar, for example).
— Attraction of pollinators: when the perianth is very small, the stamens are the only attractive parts (e.g., Mimosoideae).
— Production of fragrances: this is the case with Chloranthaceae and some Solanaceae.
— Production of nectar: in the Lauraceae, nectaries are carried by the filaments of stamens, which are external. The internal male parts are often staminodes that can also produce nectar.

The **staminodes**, stamens in which the anthers are not developed (Fig. 2.10), also have this type of function: examples are the osmophores in the

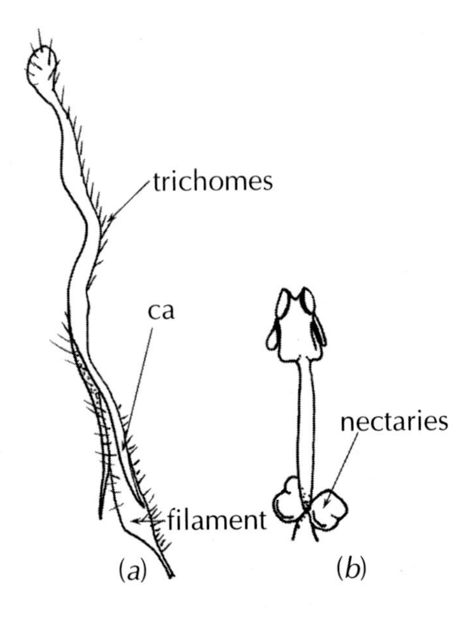

Fig. 2.9: Stamens (*a*) with trichomes of laurel-rose (Apocynaceae) and (*b*) with nectaries (Lauraceae).

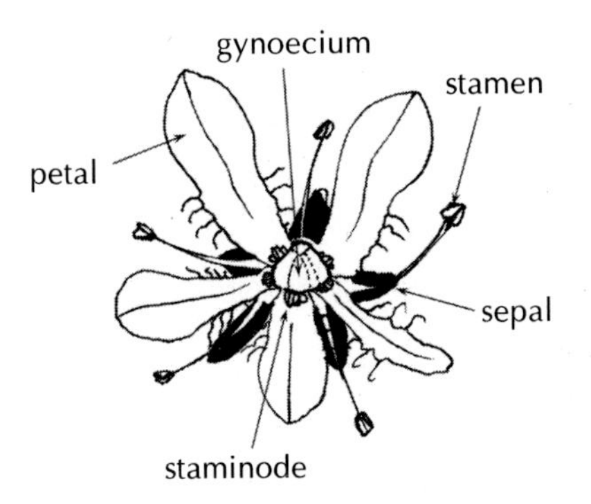

Fig. 2.10: Diagram of *Parnassia fimbriata* flower (Parnassiaceae) showing the staminodes.

Austrobaileyaceae (Endress, 1994), or visually attractive parts in some Bignoniaceae. In some cases, the staminodes do not have a known function.

c) Gynoecium

The gynoecium, a central part of the flower, is composed of one or several **carpels**. The carpels are characteristic of Angiosperms (see also the introduction) in that the ovules in them are hermetically sealed, unlike in the Gymnosperms, in which the ovules are at least temporarily in contact with the exterior.

From an evolutionary point of view, the carpels are believed to be derived from a "carpellary leaf", i.e., one containing ovules, which will close up. However, there seem to exist cases of limited, i.e., imperfect, angiospermy in the Angiosperms. This character has often been attributed particularly to members of Winteraceae such as *Degeneria*, but it is really not so. Even though the suture is clearly visible in a cross-section in this genus, the carpels are completely closed.

The gynoecium is made up of an ovary, a swollen lower part containing the ovules, an elongated style, and a uni- or multipartite stigma (Fig. 2.11).

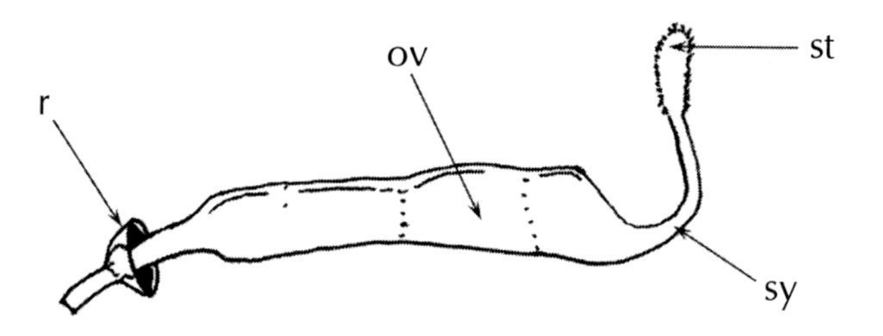

Fig. 2.11: Diagram of isolated pistil of *Lathyrus*. r, remnants of calyx; ov, ovary; sy, style; st, stigma.

Number and union of carpels

Most often, the gynoecium comprises many carpels. The carpels are united (**syncarpous** gynoecium) or free (**apocarpous** gynoecium). More than 80% of Angiosperms are syncarpous, while around 10% are apocarpous. The remaining 10% are represented by species with a monocarpellary gynoecium. In the case of free carpels, at least two types are found (Fig. 2.12).

- **ascidiate** (or ascidiform, from the Greek *askeidion*, small goatskin) carpels in which a line of closure, containing the edges of the carpel (thus delimiting the internal and external surfaces), is found at the tip of the organ;
- **sutured** carpels in which a short suture runs from the top to the bottom of the organ.

In the first case, the suture occurs at the same time that the organ develops (**congenital** suture), and we thus do not see anything on the final carpel; in the second, the suture forms after the formation of the carpellary walls (**postgenital** suture). In some intermediate cases, the suture line is not complete (**peltate** carpels).

When the carpels are united (connate), the ascidiform/sutured character reflects the manner in which the carpels have become connate: the initially ascidiform united carpels are still partly independent and are automatically separated by the septa to the extent that the suture of each carpel forms

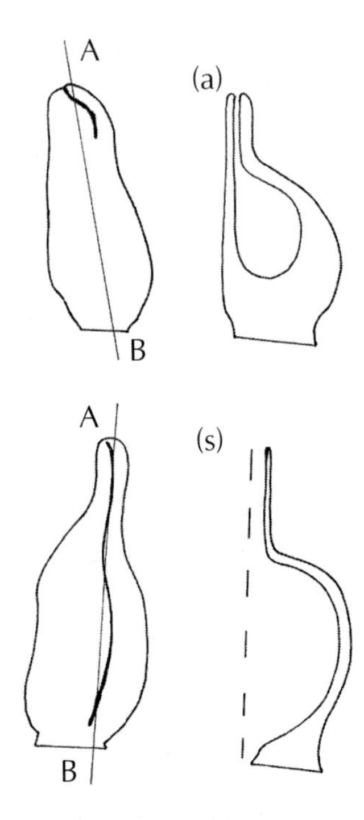

Fig. 2.12: Diagram of ascidiform carpels (a) and sutured carpels (s) seen from the exterior (left) and in longitudinal section along AB (right). The broken line below represents the position of the suture.

precociously, while a late suture following the sutured mode leads to common suture lines so that we may even end up with a single common locule at all the carpels. Septate ovaries in which the septa extend from the walls of carpels are called **syncarpous**, while if a single chamber is common to many carpels, we call it a **parasyncarpous** ovary (Fig. 2.13). There are intermediate cases in which the carpels are only partly sutured, and thus the union between different carpels is not total.

Style, stigma and transmitting tissue

Definitions

The stigma is the apical zone of the gynoecium, functionally important during fertilization since it receives the pollen grains, with some exceptions (see especially the case of **hyperstigma**, section 1.1.2a). Subsequently, the pollen grains germinate on the stigma and the resulting pollen tube reaches the transmitting tissue.

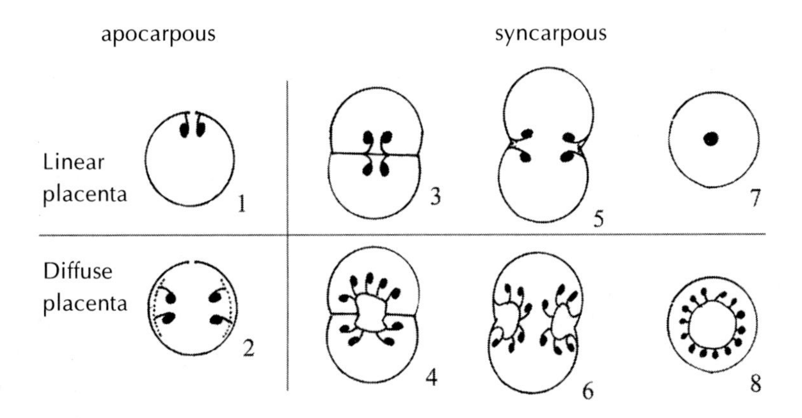

Fig. 2.13: Principal placental configurations of gynoecia seen in cross-section diagrams. Axile (3, 4), parietal (1, 5, 6), basal (7), free central (8), laminar-diffuse (2), protruding-diffuse (4, 6). The ovules are indicated with a heavy black dot. See also Fig. 2.17 (modified from Endress, 1994).

As has been said, the carpels, and thus the gynoecium, result from an organ that closes on itself and an internal and external surface are thus distinguished. The transmitting tissue is a sort of prolongation of the secretory cell zone of the stigma. The cells of the transmitting tissue develop from the first and sometimes the second cell layer of the internal surface of the carpels. Like their stigmatic homologues, they have secretory capacity. Each carpel of an apocarpous or syncarpous-ascidiform gynoecium has its transmitting tissue. On the other hand, the carpels of a syncarpous-sutured gynoecium have partly common transmitting tissue, forming a complex called the **compitum**. The compitum is better developed if there are a larger number of ovules in the ovary (in contrast, probably for reasons of feasibility, the compitum is seen only when the number of carpels is not large, frequently 5). This complexing of transmitting tissue also has an effect on the growth of pollen tubes, because it makes possible a more aggressive competition between tubes than if each carpel had its own transmitting tissue. Finally, the existence of a compitum allows a pollen tube to "change" carpels during its growth. The stigmatic secretions and those of the transmitting tissue, which are similar, contain water, galactans, arabinogalactans, ions, etc. Unlike the transmitting tissue, the stigma also secretes lipoproteins and lipopolysaccharides. The similarities in secretions as well as at the histological level have often led to the supposition that the transmitting tissue and stigma are homologous, i.e., derived from the same tissue, from an evolutionary point of view. It has been proposed that the stigma is derived from the transmitting tissue (Lloyd and Wells, 1992). This hypothesis is put forward within the context of the acquisition of the stigma by the Angiosperms, a key step that allowed control of fertilization.

Typology

Histologically, two categories of stigma are easily recognizable: stigmas of type I, called **wet**, and stigmas of type II, called **dry**. Wet stigmas have significant secretory activity, while dry stigmas have only a fine pellicle of secretion. The secretory cells belong to the epidermis only in dry stigmas and to the epidermis and hypodermis (i.e., sub-epidermal stratum) in wet stigmas.

> ***Remark:*** The secretory cells of the stigma are often more or less detached from one another, so that the stigma surface appears to have papillae and is thus called **papillate.** The cells of the stigma surface cannot always be described in this manner. For example, the cells of dry stigmas of Poaceae form **plumules**, not papillae. An extreme case of spacing between cells by secretions is that of Orchidaceae, in which the cells are completely detached. This is described as the stigmatic **cellular suspension**.

Stigmatic Secretions

A **wet** stigma produces a secretion that accumulates over the lifetime of the stigma. This is an **emulsion** that contains lipids, glycoproteins, polysaccharides, phenolic compounds, and free amino acids. **Dry** stigmas simply have an aqueous **pellicle** containing proteins. In some species, the stigma has a cuticle, and the secretion occurs below that cuticle. In both cases, the stigmatal secretion has a non-specific esterase activity, the biological significance of which is yet to be determined.

Some authors distinguish a third category, the **tubular** stigmas (Schill et al., 1985), the secretory surfaces of which are partly protected in a tube. Some associations have been proposed between the stigma types and the control of fertilization, notably between sporophytic self-incompatibility and dry stigmas, and between gametophytic incompatibility and wet stigmas (see Chapter 4). However, the correspondence is not general.

Transmitting tissue and transmitting tract

It is important to distinguish clearly between the transmitting tissue and the transmitting tract. The term *tissue* refers to cells having a secretory activity and making possible the growth of the pollen tube. On the other hand, the term *tract* refers to the medium in which this growth occurs. For example, the transmitting tract may be completely external to the cells, i.e., the extension of the ovarian chamber—or of the carpel—into the style, the **stylar canal** (also called **lumen**). This allows us to distinguish three types of transmitting tract (Fig. 2.14):

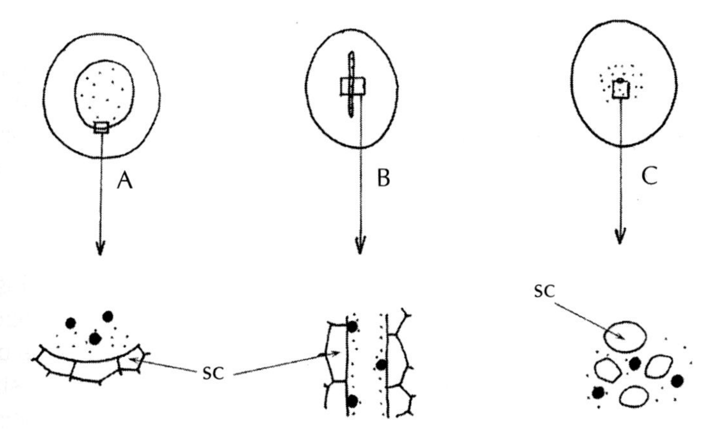

Fig. 2.14: Three types of style (top) and transmitting tract that are associated with them (bottom). (A) style with developed lumen, external tract, (B) style with narrow lumen and superficial tract, and (C) style with reduced lumen and internal tract. The tract is indicated in small dots, the passage of the pollen tube is indicated with a heavy black dot. sc: stylar cells.

— **External** tract: the lumen is wide and full of secretions of transmitting tissue.

— **Superficial** tract: the lumen is narrow and more or less twisted, offering a greater surface of contact between secretory cells and the lumen. The lumen is filled with secretions.

— **Internal** tract: the lumen is small, even imperceptible, the secretory cells, which are detached from one another, are surrounded by their secretions. This "suspension" constitutes the medium in which the pollen tube grows.

Remark: In the first two cases, the secreted gel is the growth medium for the pollen tube.

Specific guides of the pollen tube

The transmitting tract constitutes, as we have said, the medium of progression of the pollen tube. At the end of the transmitting tract, the tube opens into the carpellary (or ovarian) chamber and reaches an ovule. However, this step may be considerably facilitated, particularly when the transmitting tract comes into direct contact with the micropyle or nucellus (Fig. 2.15). In the Euphorbiaceae, the nucellus is in direct contact with the placenta region of the transmitting tract, by means of a protuberance, the **obturator**. Obturators also exist in certain Liliaceae (Endress, 1994). The placenta may also have an extension penetrating the lumen directly (e.g., Lentibulariaceae). In the Anacardiaceae (mango, pistachio), the anatropous ovules are in contact with the transmitting tract by a protuberance of the funiculus.

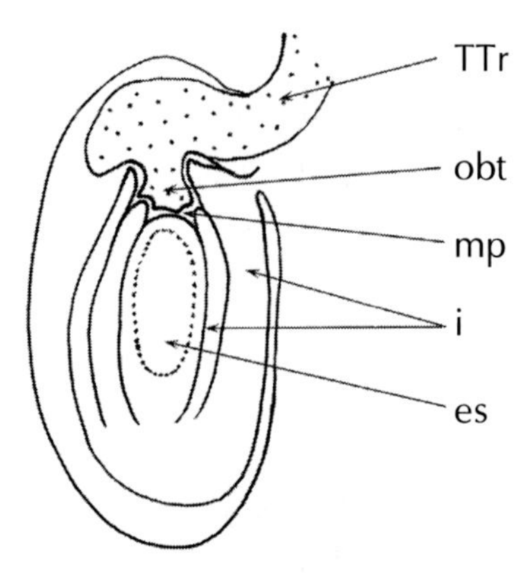

Fig. 2.15: An example of specific guides of pollen tube: obturator of Euphorbiaceae. TTr, transmitting tract; obt, obturator; mp, micropyle; i, ovular integuments; es, embryo sac (modified from Endress, 1994).

Remark: The pollen tube may also progress in non-specialized structures. This is especially true of certain **cleistogamous flowers** of Malphigiaceae (*Janusia, Gaudichaudia*): while the floral bud is closed, the pollen grain germinates in the anther and the pollen tube grows through the filament, reaches the receptacle, and penetrates the gynoecium. In cleistogamous flowers of other families, the pollen tube pierces the anther to reach the stigma directly.

Ovule and placenta

Definitions

Ovules are structures containing the female gametophyte or **embryo sac** (Fig. 2.16). The embryo sac is surrounded by a diploid tissue, the **nucellus**, which

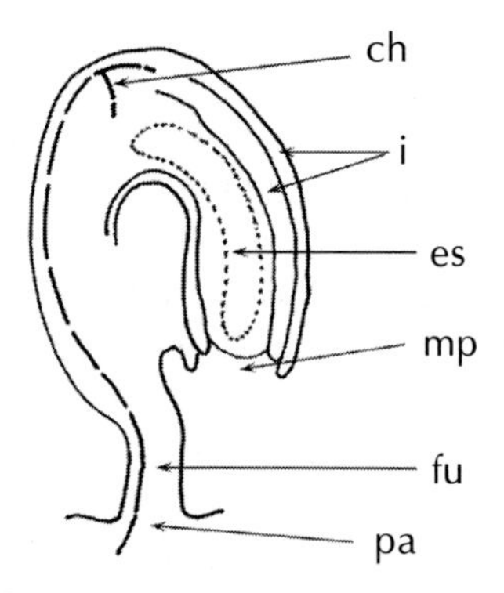

Fig. 2.16: General structure of the ovule: example of a campylotropous ovule of Fabaceae. ch, chalaza; i, integuments; es, embryo sac; mp, micropyle; fu, funiculus; pa, placenta. The conducting elements are indicated in broken lines.

is itself enveloped in two **integuments** (but not always, see below). The integuments make an orifice, the **micropyle**.

The ovule is attached to the carpel wall or more specifically to its zone of fixation to the carpel, the **placenta**, by the **funiculus**. A cribro-vascular bundle traverses the funiculus and ends in a terminal branch located in the region opposite the micropyle: the **chalaza**.

Typology of placentas

The eusyncarpous gynoecia bear their ovules on the carpellary walls towards the central axis; this is why it is called **axile** placentation (e.g., tomato, Solanaceae). The parasyncarpous gynoecia have a **parietal** placentation, where the ovules are arranged on the flanks of the carpels, often along the sutures (e.g., Brassicaceae). Finally, the ovules of parasyncarpous gynoecia may be arranged in a central placental column originating from a dilatation of the base of the ovarian chamber. This is why the term **basal** placentation is used (e.g., Primulaceae). In some rare cases (Combretaceae), the inverse is found: the placental column begins at the summit of the ovarian chamber, and this is called **apical** placentation (Fig. 2.17).

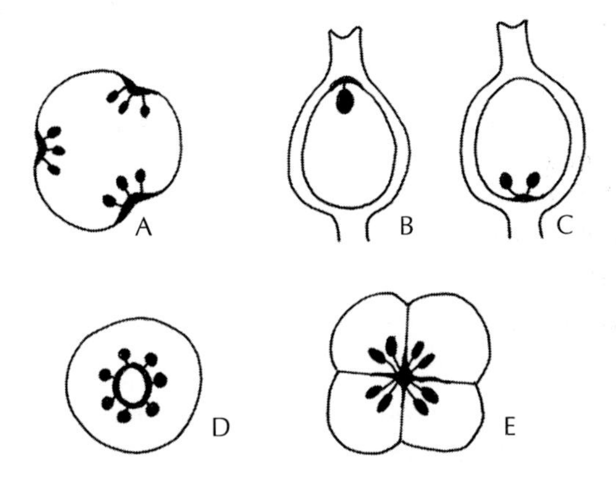

Fig. 2.17: Different modes of placentation: parietal (A), apical (B), basal (C), axial or free central (D), and axile (E). See also Fig. 2.13.

The ovules of eusyncarpous, parasyncarpous, and apocarpous gynoecia are frequently aligned, along **linear** placentas, or distributed over a more or less protuberant surface, in which case they are called **diffuse** placentas. The basal placentas are automatically diffuse. In contrast, the axile and parietal placentas are diffuse only if they are dilated. In this case, these placentas are called **protruding-diffuse** (e.g., Passifloraceae). On the other hand, the placentae of carpels of apocarpous gynoecia are not in principle protruding and the ovules are arranged on several lines rather than a single one, towards the flanks of the carpel (on either side of the line of suture). These are the **laminar-diffuse** placentas (e.g., some Nympheaceae, Hydrocharitaceae).

Ovules

Even though all ovules are constructed in the same way (integuments, nucellus, embryo sac), three main forms of ovules are found (Fig. 2.18):

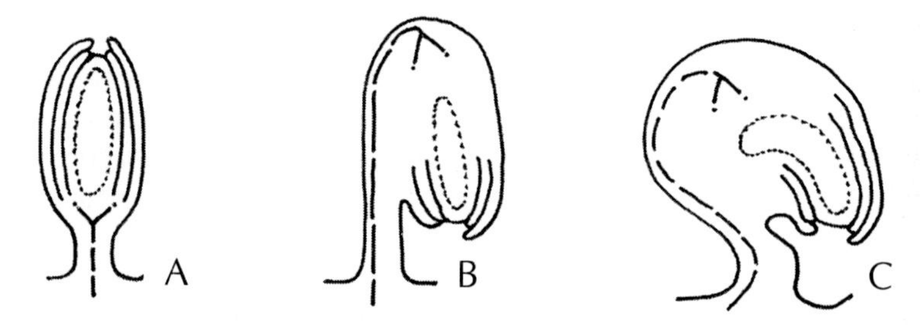

Fig. 2.18: Ovular configurations: orthotropous (A), anatropous (B), and campylotropous (C). The indications are the same as in Fig. 2.16.

- **Orthotropous** ovules: The ovule is erect, the micropyle being opposite to the funiculus. This type is found only rarely in the Angiosperms.
- **Anatropous** ovules: The ovule is curved (because one side grows more than the other, see below), so that the micropyle is close to the funiculus, the chalaza being opposite to the funiculus. The cribro-vascular bundle creates a relief on the surface of the ovule, the **raphe** (e.g., Corylaceae).
- **Campylotropous** ovules: The ovule is curved to such an extent that the curvature affects the embryo sac, which is not the case in the anatropous ovules. This bending of the embryo sac makes the sac generally longer in this type of ovule (e.g., Fabaceae).

Ovules are often described on the basis of observations related to curvatures. However, there are other characters very important from the evolutionary perspective, notably the number of integuments and the number of cell layers of the nucellus surrounding the embryo sac:

- **Bitegmic** (or **dichlamydeous**) ovules constitute the majority of ovules of Angiosperms. However, there are **unitegmic** (or **monochlamydeous)** ovules, the evolutionary significance of which will be discussed later in this book.
- The number of nucellar cell layers surrounding the embryo sac is greater than 1 (**crassinucellate**) or equal to 1 (**tenuinucellate**).

Remark: Exceptionally, there are ovules with **three integuments** (some Annonaceae) or ovules that are **ategmic** (some Santalaceae, Gentianaceae).

The most frequent case is that of anatropous-crassinucellate-bitegmic ovules. This is present in the majority of Magnoliidae and Rosidae. Another common case is that of anatropous-tenuinucellate-unitegmic ovules, as with Asteraceae and Orchidaceae.

Orthotropous ovules constitute an interesting case in relation with their arrangement in the carpels. It is in fact supposed that the **curvature** of the anatropous or campylotropous ovules constitutes an **adaptive advantage** during sexual reproduction, because the pollen tube frequently reaches the placenta and, when there is an ovular curvature, finds the micropyle easily, since the micropyle is turned towards the funiculus. This advantageous arrangement does not exist in flowers with orthotropous ovules: thus, some species have developed **alternative strategies**. For example, in Flacourtiaceae, the relatively elongated ovules are not perpendicular to the placenta, so that the micropyle is nearer to the placenta of the next ovule. In pepper (*Piper* sp., Piperaceae), a single very large ovule exists in the carpel, the micropyle of which is very close to the end of the transmitting tract. In some Araceae (*Pistia* sp.), the placentation is basal so that the micropyles are all turned upward. The ovarian chamber is filled with a secretion identical to that of the transmitting tract. All these strategies considerably increase the probability of success of the pollen tube in its "search" for the micropyle.

d) Bracts

In the introduction it was indicated that one definition of *flower* emphasizes the importance of the bract, which is found at its base. It is true that the flowers of Angiosperms generally have a bract found at the base of the floral peduncle, where there is a peduncle. However, sometimes there are no bracts. The shape of the bract is highly variable, as is its involvement in biological functions (Fig. 2.19). Protection of the flower is one of its most obvious roles. An extreme case is that of Poaceae, in which spikes of two or three flowers (in general) are protected by a pair of bracts (glumes) and flowers by bracteoles (lemma and palea). However, even though the latter are considered similar to bracts, they are not so from an embryological point of view: lemma and palea serve as sepals. Lodicules also do not serve as inner bracts but as petals (see below). The bracts may be the only organs that serve to protect the flower: this is the case with Passifloraceae, in which sepals and petals are attractive. The bracts protect the flower, especially in the Heliconiaceae and the Musaceae, even when it is closed, forming a sort of cone from which only part of the flower protrudes.

A remarkable case is that of **attractive** bracts: in many groups, the bract helps attract pollinators and is thus larger and coloured. In many species, the bracts are vividly coloured (e.g., *Castilleja miniata*, Scrophulariaceae).

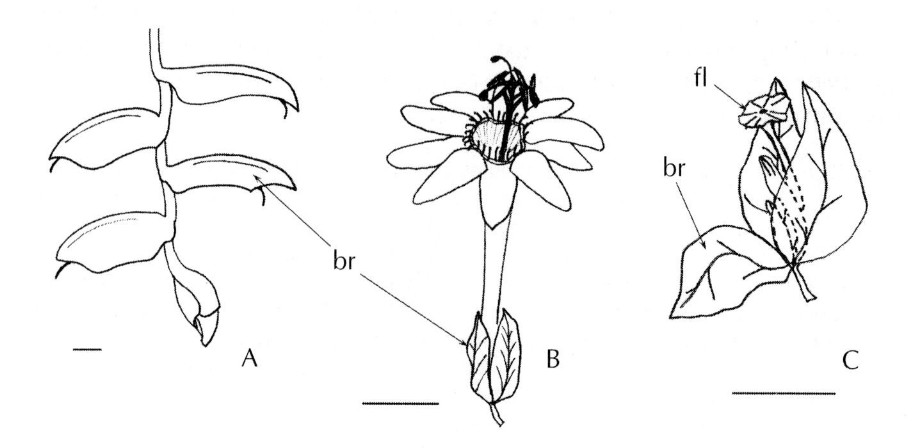

Fig. 2.19: Three examples of bracts: protecting bracts in an inflorescence of *Heliconia* (A), basal bracts in *Passiflora* (B), and attractive bracts of *Bougainvillea* (C). In *Bougainvillea*, three flowers are grouped together, each having a vividly coloured bract. One bract has been turned down in order to reveal the cyme. br, bract; fl, flower. The bar represents around 1 cm.

When the flowers are arranged in a dense inflorescence, such as capitula (see Chapter 1), the bracts sometimes all extend from the base of the inflorescence, in the form of a collar or **involucre**. This is especially the case with Asteraceae. On the other hand, all dense inflorescences do not have this characteristic, as with Dipsacaceae (teasel, scabious, etc.). This constitutes the difference between **capitulum** and **head**.

2.1.3. Formalization of the Floral Structure

a) Floral formula

The description of the floral structure can also be written as a **floral formula** or **floral equation**. There are two ways of symbolizing the different floral organs or the relationships between them: one is French and the other is American. The latter is adopted here (see Judd et al., 1999) because it presents a complete codification covering nearly all aspects.

The symbols used for sepals, petals, stamens, and carpels are respectively *S, P, E,* and *C*. The number of each organ is indicated first and the whole is linked together with commas: e.g., *5 S, 5 P, 10 E, 3 C* or even *5, 5, 10, 3.* No particular mention is made when the floral organs are free; on the other hand, when floral organs of an identical nature are united (**connate**) the term concerned is circled and when organs of different nature are united (**adnate**) the two types of organs are connected by an arc. The lines are broken when the union is incomplete. Carpels are marked with a line above when the ovary is inferior and underlined when the ovary is superior. Sometimes, the number of organs of a type fluctuate between two extreme values, which is noted for example as *5-7 P*. The number of organs may be indeterminate, in which case it is called infinite (indicated by the symbol ∞), or even null, in which case it is written as *0*. A whorl of stamens or carpels may be composite and comprise fertile and sterile organs (**staminodes** and **pistillodes**, respectively), the latter being indicated by a point: e.g., *5C*• + *5C*. The organs that are morphologically in between petals and sepals (tepals) are indicated between hyphens: *-5-*. Finally, there are two additional pieces of information: the symmetry (* signifies actinomorphy and *X* zygomorphy) is written before the floral formula and the type of fruit is written in full after the formula. The example shown in Fig. 2.20 summarizes the principal symbols.

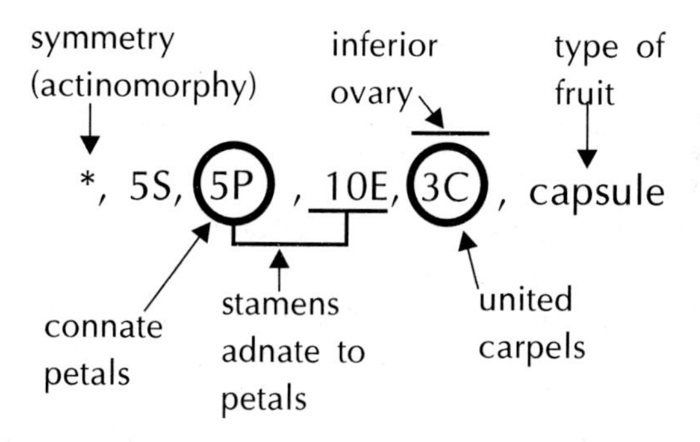

Fig. 2.20: An example of floral formula recapitulating the principal characters of the floral structure.

b) Floral diagram

Flowers can also be represented in **floral diagrams**. These are said to express what is seen when a cross-section of a floral bud is made. The whole is "stylized" to the extent that the plane of the section goes through all the whorls. Moreover, the diagram includes all the very important information about the floral architecture, including unions. The bract is represented by a sort of triangle, and the inflorescence axis by a small circle opposite to it. Sepals are indicated by an open crescent, petals by a closed crescent, and stamens by a small ellipse or by the appearance of a cross-section of a stamen (this option demonstrates more clearly whether the dehiscence of the stamens is extrorse, introrse, or lateral). Finally, carpels are indicated by a compartment with a form that is in agreement with that observed in the flower and delimited by a double line or thick line. True septa are generally indicated in double or thick lines, false septa in a single line (the distinction between true and false septa is not always apparent on the diagrams). When the septa have disappeared secondarily, a notch is nevertheless marked at the position they would occupy, in order to show the zones of union of carpels. Ovules are indicated by a black dot, indicating their position in the flower. In some cases, the floral diagrams must be modified because of the presence of spurs, etc. **Nectaries** in particular are indicated by a small swelling of the organs that support them. **Unions** are represented by dotted lines between the organs that are united. Sometimes, a distinction is made between incomplete union (dotted lines) and complete union (solid lines).

In general, organs supposed to be absent or being a secondary loss observed in the species considered are indicated by a cross. Moreover, there is a kind of "convention" of position that reflects the reality:

- The bract is placed opposite to a petal in Dicotyledons, except in Leguminosae (Fabaceae), or more generally diametrically opposite to a sepal.
- The inverse is true in Monocotyledons, except Orchidaceae: the bract is opposite to a sepal.
- The bract is indicated at the bottom of the diagram, the inflorescence axis at the top.

A sample floral diagram, of a tulip (Liliaceae), is given in Fig. 2.21. The sepals and petals (respectively external and internal tepals) can be indicated by the same symbol—an empty crescent—in order to emphasize that they serve as tepals, but it is preferable to conserve the sepal/petal distinction, since the nature of the tepal is moreover noted in the floral formula. In other cases, it happens that the parts of the perianth do not have a perfectly straight arrangement but are imbricate, etc., which must be indicated.

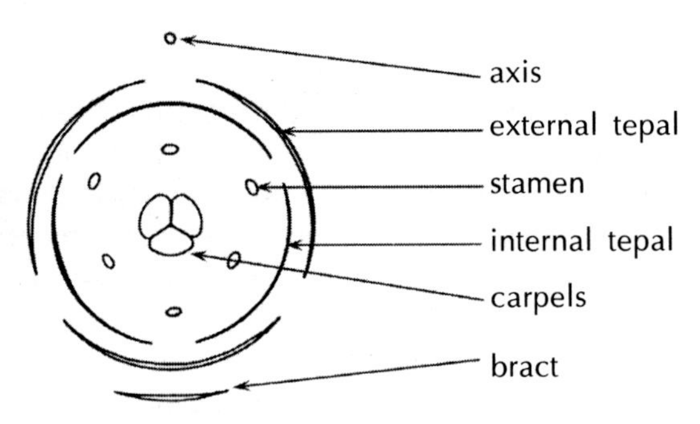

Fig. 2.21: An example of floral diagram: a tulip.

2.2. GENETIC EXPRESSION AND MORPHOGENESIS

This section will be limited to an explanation of the morphogenesis of the floral architecture as a whole; the aspects of development at the histological scale or of different floral organs are not addressed (for a review, see Vallade, 1999, and Schneitz et al., 1998). The objective is to discern the elements of genetic control that will allow us to understand subsequently the evolution of Angiosperms through the MADS genes in particular.

The genetic aspects that are addressed occur at the apical meristem that has undergone **floral evocation** (for details of the physiological order, see Taiz and Zeiger, 2000). This meristem is structured in three parts: the superficial cell layer with anticline divisions (L1), the cell layer that underlies it (L2), and finally the internal cell mass (L3). These different layers are the source of different parts of floral organs: for example, L1 forms the epidermis (see Vallade, 1999).

2.2.1. Model of Genetic Switching

The genetic aspects of floral development have been considerably refined these past years, particularly through the study of floral mutations of *Arabidopsis thaliana*. The studies of Meyerowitz et al. on **homeotic** mutants have shown the existence of a system of genetic switching, designated as the "ABC system", responsible for the establishment of floral whorls. Three categories of mutants (Fig. 2.22) are of particular interest: *apetala-1, apetala-3, agamous*, and *pistillata* (hereafter denoted by *ap-1, ap-3, ag*, and *pi*). In the *ap-1* mutants, it is observed that there are no sepals or petals, and instead there are carpels and stamens (structure of type *C, E, E, C*). In the *ap-3* or *pi* mutants, or

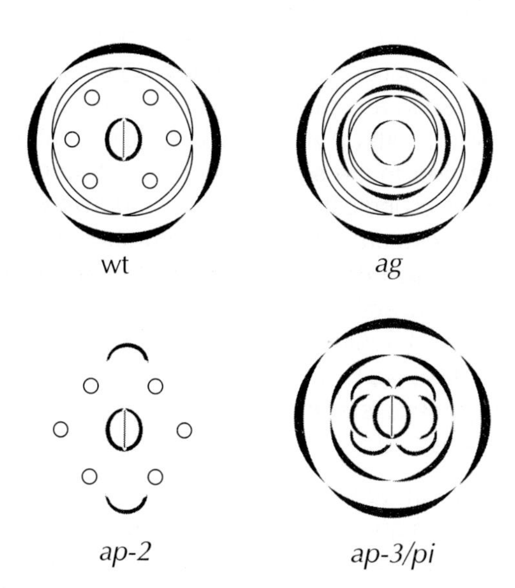

Fig. 2.22: Simplified floral diagrams of wild type (wt) and *apetala-3* or *pistillata* (*ap-3/pi*), *apetala-2* (*ap-2*), and *agamous* (*ag*) mutants of *Arabidopsis thaliana*. The adaxial side is turned toward the top.

even in the double *ap-3/pi* mutants, there is no petal or stamen, and there is a construction of whorls of the type *S, S, C, C*. The *ag* mutants have only sterile parts; the structure is of the type *S, P, P, S*. In the second type of mutants, function seems to have disappeared at the level of the second and third whorls, which are affected only morphologically. In the wild type this would allow the expression of petals and stamens. However, when this function does not exist, the floral organs are not absent, but only inadequate (there is thus a **homeotic** mutation). Moreover, mutants of the first and third types are symmetrical (i.e., with a *x, y, y, x* structure) and complementary (the organs of *ap-1* are precisely those lacking in *ag* and vice versa), which suggests that there are two genes with antagonistic functions expressed respectively in the whorls 1 + 2 and 3 + 4. From these observations a system can be elaborated with three "functions" A, B, and C, determined respectively by the genes *ap-1, ap-3/pi*, and *ag*. The combination of *ap-1* and *ap-3/pi* forms petals, the combination of *ap-3/pi* and *ag* forms stamens. The expression of *ap-1* alone forms sepals, and the expression of *ag* alone leads to carpels. The equivalents (homologues) in *Antirrhinum majus* of four genes of *Arabidopsis thaliana* are respectively *squamosa* (*squa*), *deficiens* (*def*)/*globosa* (*glo*), and *plena* (*ple*).

> ***Remark:*** However, although the genetic switching model has been explained provides a convincing explanation about the determination of the establishment of the floral architecture, some problems do remain. For example, mutants B and C of *Arabidopsis* and *Antirrhinum* differ slightly at the phenotype level. But the problems are greater with respect to function A: in *Arabidopsis*, function A covers in fact 2 genes: *ap-1* and *ap-2*. The genes *ap-1* and *squa* are relatively similar structurally and functionally (this is not the case with *ap-2*) but the correspondent mutants also have different phenotypes. In the *squa* mutants, the flowers develop into inflorescence axes, which indicates that this gene is involved in the floral identity of the apical meristem. An action comparable to the *ap-1* in the gene *squa* has not yet been demonstrated. Even in *Arabidopsis*, the mutant *ap 1* does not have exactly a structure of the *C, E, E, C* type predicted by the model. Function A seems to be required in order to determine the floral meristems thus destined to produce sepals. Functions B and C are added to that to specify petals, stamens, and carpels. Function A involves in fact three types of genes: *ap-1, cal, ful* (see also Chapter 1).

The ABC system that we will examine is at present complemented by two other functions, the existence of which is suggested by the observation of results of various genetic manipulations. In the B mutants (*ap-3/pi*), there are certainly carpels in whorl 3, in *Arabidopsis* as well as in *Antirrhinum*, but the ectopic expression of C genes (under the control of promoter CMV 35S inducing a constitutive expression) does not form carpels on distinct regions of floral structures: outside the flower, the alterations observed concern a curvature of leaves or a slight modification of the bract. Similar observations have been made with functions B and A. The effect of ABC genes is thus possible only if there are one or more additional factors specific to the flower. In particular, experiments of co-suppression of *FBP2* (the introduction of copies of the gene extinguishes its own expression as well as that of the endogenous gene; this is called *silencing*) in petunia or expression of *TM5* antisense in tomato, both MADS genes, produce homeotic changes in the three internal whorls and loss of floral determination. There are supplementary genes, called intermediates (hence *Im*), the importance of which is emphasized in flower development. In phylogenetic terms, these *Im* genes are arranged in a paraphyletic group, which corresponds to the "Im function" (Fig. 2.22).

Recently, studies in *Arabidopsis* have elucidated more precisely the role of factors Agl2, Agl4, and Agl9 (Pelaz et al., 2000). The corresponding genes are known to be expressed before genes *ag, pi*, and *ap-3* in whorls 2, 3, and 4. The monogenic mutations *agl2, agl4*, or *agl9* have practically no effect on the floral morphology, while the triple mutation *agl2 agl4 agl9* led to a homeotic phenotype in which the 3 innermost whorls are replaced by sepals. The genes *agl2, 4*, and *9* act in parallel with *ag, pi*, and *ap-3* and are neither downstream nor upstream of the genetic regulation pathway of functions B and C. In fact, the expression of *ag, pi*, and *ap-3* is not affected by the triple mutation *agl2 agl4 agl9*, just as the double mutations *ag pi* and *ag ap-3* do not affect the expression of *agl2 agl4* and *agl9*. Thus, the genes *agl2, 4*, and *9*, which have been renamed *sepallata 1, 2*, and *3*, determine a genetic function acting on the whorls of petals, stamens, and carpels, which is called "**function E**".

Although *in vitro* experiments have demonstrated that Pi and Ap-3 are associated (heterodimerization) and that two Ag proteins are associated (homodimerization), the interactions between the ABC and Im factors are yet to be precisely understood. Experiments of double hybridization have revealed interactions not between Ple proteins (equivalents of Ag in the snapdragon) but between Ple and other factors (MADS) DEFH200 and DEFH72 (Davies et al., 1998) belonging to the group of Im factors. Still, in *Antirrhinum*, it has been shown through experiments using the ternary factor trap technique that it is the heterodimer Def-Glo that interacts with factors Im DEFH200 and DEFH72 (Gutierrez-Cortines and Davies, 2000). Finally, in *Arabidopsis*, it seems that Agl9 interacts with Ap-1, and that Ag interacts with Agl2, 4, and 9 (Pelaz et al., 2000).

Remark: Through the ternary factor trap technique, an interaction can also be shown between Squa and the heterodimer Def-Glo. Function A could thus be implicated not only in the determination of whorls 1 and 2, but also in that of whorl 3.

Figure 2.23 presents models of establishment of floral architecture. Part 1 shows the basic model of the 1990s and part 2 shows a simplified version of the present model. It is to be noted in particular that there seems to exist a function D responsible for the setting of ovules in the carpels. Indeed, the "ABC" system of genetic switching that we have mentioned has been extended to a **function D**, determining the carpels in petunia and coded by genes *FBP7* and *FBP11*, the ectopic expression of which causes the formation of ovules on the perianth (see also section 2.2.6). Genes of group D were initially studied in this plant, and a single equivalent has so far been identified in *Arabidopsis thaliana*: *AGL11*.

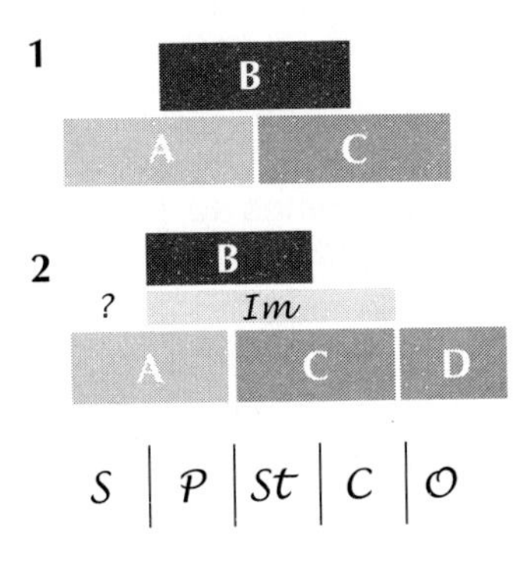

Fig. 2.23: Genetic determination of the setting of whorls in *Arabidopsis thaliana*. (1) Basic model of the 1990s. (2) Present simplified model. S, sepals; P, petals; St, stamens; C, carpels; O, ovules.

Finally, it should be stressed that genes for functions A, B, C, D, and E in *Arabidopsis* act in concert with non-homeotic genes. However, the present-day understanding of this subject is fragmentary and the modalities of action of these genes in the determination of floral identity remain to be precisely described, which is why we refer strictly to the experimental data.

— The mutants *spt* (*spatula*) and *crc* (*crabs claw*) show altered gynoecium development. Moreover, in the triple mutant *ag ap-2 pi*, the floral organs present some structures of stigmatal appearance, which do not exist in the quintuple mutant *ag ap-2 pi spt crc* (Alvarez and Smyth, 1999). Thus, Crc and Spt seem to act independently of Ag to fix carpel identity.

— The phenotype of weak mutations of *ag* (the third whorl still forms stamens) is accentuated by the introduction of a double mutation on *hua1* and *hua2* (Chen and Meyerowitz, 1999). In contrast, the simple mutants *hua1* and *hua2* do not have a floral phenotype and the double mutation *hua1 hua2* alone does not suffice to completely change the identity of floral organs on whorls 3 and 4. The mutation *hen1* accentuates the phenotype of double mutant *hua1 hua2*, with a conversion of stamens into petals. The triple mutant *hen2 hua1 hua2* presents petaloid or even sepaloid parts on the third whorl, with a transcription of *pi* and *ap-3* weakened in whorl 3, at least at the beginning of development. Besides, in the triple mutants *hen1 hua1 hua2* and *hen2 hua1 hua2*, the transcription of *ag* is normal in whorls 3 and 4 at the beginning of development but does not subsequently reach the level observed in the wild type. It is thus possible that *hua1* and *2* act in synergy with *hen1* and *2*, downstream of *ag*, after a bifurcation of the genetic regulation pathway triggered by *ag*; this is corroborated by the fact that the strong mutations *ag* do not change the phenotype after introduction of supplementary mutations *hua1* and *hua2* (Jack, 2002). Moreover, these genes could exert a positive feedback on *ag* expression, as attested by transcription levels.

2.2.2. Upstream of ABC (D)

As we have mentioned, some genes such as *ap-1* have an ambiguous function because they determine the floral identity of the primordium as well as the identity of certain floral organs. In addition, genes such as *leafy* acting upstream of the "ABCD" genes are also involved in the fixation of the floral identity of the primordium. How do these genes of floral identity trigger the homeotic genes ABCD?

The **gene leafy** (*lfy*) seems to be an important link in this process. The mutants of *leafy* deletion have sets of parts with the appearance of small leaves, instead of a flower, and show a very small expression of *ap-3* and *pi*. The expression of *ag* is only retarded and the expression of *ap-1* is nearly normal. These data suggest that *lfy* is a gene that promotes the expression (particularly as transcription activator) of genes ABC, or even that it acts in the whorls to specify a "floral" identity, so that, when not expressed, the whorls produce leaf-type organs. Finally, *lfy* may have two simultaneous roles: identification of the floral meristem and activation of homeotic genes. Some experiments conducted on yeast show that *lfy* is a transcriptional activator in which the targets are not known *in vitro*. The mutants of *Arabidopsis* expressing *lfy* constitutively show flowers of the *C, E, E, C* type, with a reduction of the number of carpels in whorl 4, or even entirely carpelloid flowers. Moreover, in these mutants, the expression of *ap-1* is unchanged or merely increased, but the expression of *ag* is more precocious, more extended, and more elevated than in the wild type. The expression of *ap-3* is not perceptibly modified. It seems thus that the Lfy factor is upstream of the Ag factor and could thus be a transcriptional activator of this gene. *lfy* is also weakly expressed in the leaf primordia in the wild type, whereas overexpression of *lfy* also causes an expression of *ap-1* in the leaf buds (Parcy et al., 1998). This observation suggests that *lfy* has a role in the activation of *ap-1* outside the context of floral morphogenesis. In contrast, *ag* is not generally induced in these conditions. However, it happens (in a similar genetic context) that stigmatal structures are observed on the leaves, accompanied by the expression of *ag*. This is more marked when the mutants present a defective gene *ap-2* (*ap-2* is part of the A gene group and is thus an inhibitor of *ag*).

It seems clear that *lfy* activates *ag* in whorls 3 and 4 and not in whorls 1 and 2, because of repression by *ap-2*. The system is, however, more complex because the mutants *ap-2* still have a reduced expression of *ag* in whorls 1 and 2, and because overexpressed *ap-2* does not totally extinguish the expression of *ag* in whorls 3 and 4. In particular, the group A gene called *lug* (for *leunig*), present in whorls 1 and 2, has no function of identity of floral organs but is a cadastral gene (i.e., for determination of regions); given the existence of the expression (weak) of *ap-2* in whorls 3 and 4, it is possible that *lug* helps to suppress *ag* in whorls 1 and 2 (Howell, 1999).

We have seen that the B genes (*ap-3* and *pi*) are not affected by overexpression of *lfy*. In contrast, an ectopic expression of these genes is observed in mutants overexpressing *ufo* (*unusual floral organs*). Moreover, in mutants that are null for *lfy*, deletion or overexpression of *ufo* has no effect. This suggests that *lfy* and *ufo* act in synergy at whorls 2 and 3, which is corroborated by their simultaneous expression in these whorls (Parcy et al., 1998).

It has moreover been shown that the *ask-1* mutation amplifies the phenotype of weak mutants *lfy* and *ap-3* (i.e., the phenotypic expression of the *lfy* and *ap-3* mutants is moderated), suggesting that the Ask-1 factor interacts with Lfy and Ap-3 at whorls 2 and 3. It is now supposed that Ask-1 acts in synergy with Ufo to inhibit a repressor of *ap-3* and *pi*, perhaps by ubiquitination and proteolysis (Zhao et al., 2001).

Thus, *lfy* certainly seems to be a very important link in the floral determination:

— in fixing the floral identity of the primordium and
— in activating the transcription of *ap-1* perhaps directly, of *ap-3* (and *pi*?) by interacting with *ufo*, and of *ag* by means of an unknown factor.

We have, through the extensively studied examples of *lfy* and *lug*, emphasized the existence of two types of genes intervening upstream of the **homeotic** genes (i.e., fixing the identity of floral organs and thus the mutation leads phenotypically to a mutation said to be "homeotic"): genes of **meristematic identity** (GMI) and **cadastral** genes. *lfy* is not the only GMI: there are similarly *cal* (for *cauliflower*) and *tfl-1* (for *terminal flower*); to this can be added, as has been suggested, *ap-1* and *ap-2*. By their capacity to fix a region of expression (antagonistic genes), *ag* and *ap-2* are cadastral genes. To these are added the genes *lug, sup* (*superman*) and *bel-2* (*bell*). This group is summarized in Fig. 2.24.

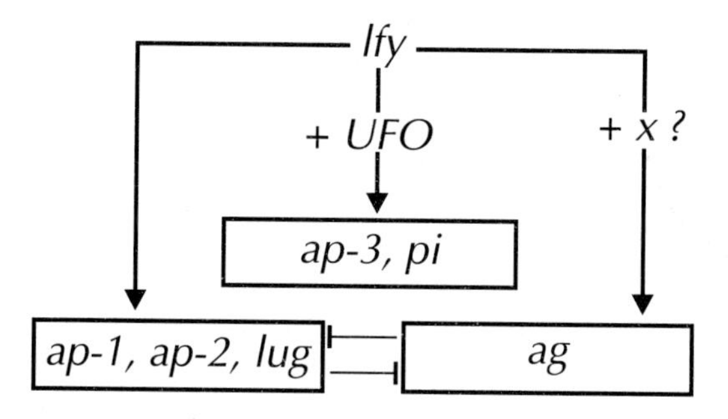

Fig. 2.24: Genetic interactions during the setting of whorls in *Arabidopsis thaliana*. Arrows indicate activation and lines indicate inhibition.

2.2.3. Fine Regulation of the Expression of ABC Genes

The recent dissection of mechanisms that regulate the expression of ABC (D) genes has allowed us to discover some subtleties. In mutants that have an ectopic transcription of *ap-3* (and *pi*), whorl 1 is not constituted of petals, contrary to what the model predicts. It has been proposed that there is also a post-transcriptional control negatively regulating *ap-3* in the first whorl (Howell, 1998). Moreover, it has been found that the factor Ap-1 was the substrate of a farnesyl-transferase, and that in the farnesylated state Ap-1 is inactive (Yalovsky et al., 2000). Even though the modalities of regulation of the enzyme remain to be specified, it is possible that Lfy has the capacity to inhibit the farnesyl-transferase and that Ag activates it. With respect to whorl 2, it seems clear that other genes such as *ant* act in the determination of the initiation of petals. Curiously, Ant is also known as a factor needed for the normal development of the ovule (see section 2.2.6). Ant seems in fact to repress the expression of *ag* in whorl 2 and, independently, acts in synergy with *ap-2* in petal development (Krizek et al., 2000). Finally, the expression of *ag* in whorls 3 and 4 is not regulated in the same way, since it seems clear that *ag* is repressed by *ap-3* in whorl 3, while it is activated by *lug* in whorl 4 (Deyholos and Sieburth, 2000).

2.2.4. Triggering the Genetic Interactions

How are the expressions of genes of floral identity of the meristem triggered? The genes whose actions are located upstream of GMI are part of the system of integration of environmental signals, such as day length. Far from wanting to tackle the determination of flower setting, we will limit ourselves to indicating that, when the signals coming from the environment (mostly), especially photoperiod, are adequate, a stimulus, probably a gibberellin, is produced in the leaves and reaches the apex. The day length is detected by the phytochrome system, the principal component of which is *phyB*. Depending on the species, it may or may not be necessary that this stimulus be maintained for a fairly long duration (for a review, see Hempel et al., 2000).

This stimulus helps set up the launching of the genetic system causing the expression of GMIs. This system is composed of three networks in *Arabidopsis*: one called autonomous (Devlin and Kay, 2000), which is controlled by temperature, one linked to photoperiodism, and the last independent of day length, which also works through gibberellins. The first network is called "autonomous" because in it the flowering constitutes a default behaviour, continually inhibited when the flowering conditions are not met. At present, the key gene of this network is *FLC* (*flowering locus C*), which is responsible for the inhibition of GMI activators. The low temperatures inhibit the expression of the *FLC* gene, promoting flowering. The search for actors of the network linked to the photoperiod led to the discovery of *co* (*constant*), for which the mutant does not flower in long day conditions. It seems clear that this gene is the origin of several divergent pathways of flowering activation (Fig. 2.25).

Finally, it must be noted that the three networks are not completely independent, and that the *co* gene in particular may lead to the activation of genes of the autonomous network.

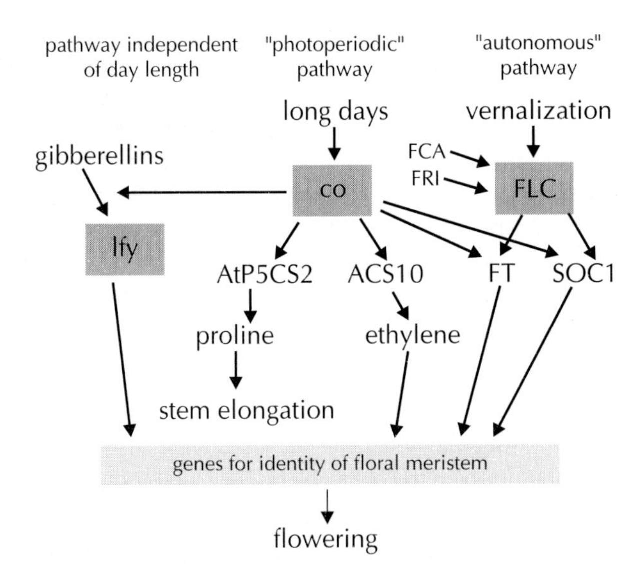

Fig. 2.25: Determination of the triggering of genes of floral identity, indicating the central role of the genes *constans* (co), *flowering locus c* (FLC), and *leafy* (lfy). See the text for details (modified from Devlin and Kay, 2000).

2.2.5. Inflorescences and the System of Genetic Switching

The transition from the vegetative to the reproductive state of the apical meristem occurs, as has been stated, by the expression of GMI. In the case of an isolated flower, this process seems clear. But what happens when there is an inflorescence, and a part of the meristem must still develop the inflorescence axis once the flower is formed? In *Arabidopsis*, *tfl* mutants produce a terminal flower, so that the growth of the inflorescence is not indeterminate. Thus, the Tfl factor seems to maintain the terminal meristem in an indeterminate state, promoting indefinite growth. Mutants ectopically expressing Tfl and Lfy/Ap-1 do not respond to Lfy or to Ap-1, which suggests that Tfl obstructs the activity of factors Lfy and Ap-1. Curiously, Lfy and Ap-1 seem to repress the expression of *tfl*, because *lfy* mutants produce Tfl in meristems of order II, which are considered to produce the flowers, and because mutants overexpressing Lfy do not produce Tfl (Hempel et al., 2000). The feedback pathways of Tfl and Lfy factors are still uncertain. Moreover, it is possible that *tfl* is not the only gene to ensure the proper functioning of the terminal meristem (Fig. 2.26).

A spectacular example of aberration in inflorescence meristem development is that of cabbage *Brassica oleracea* (cauliflower, broccoli). In these plants, a non-sense mutation in exon 5 of the *cal* gene seems to be responsible for the

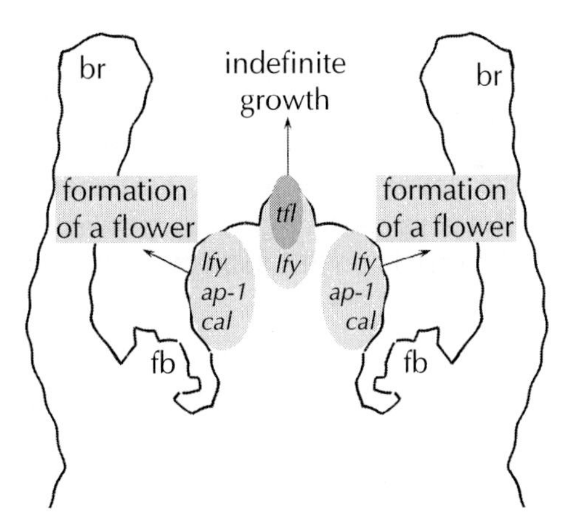

Fig. 2.26: Simplified model for genetic determination of the development of a cluster-type inflorescence at the apical bud. The genes indicated are those characterized in *Arabidopsis thaliana*: *tfl* (*terminal flower*), *ap-1* (*apetala-1*), *lfy* (*leafy*), and *cal* (*cauliflower*). The floral buds (fb) and the future bracts (br) are also indicated. The gene *lfy* is expressed widely and activates the transcription of *ap-1* and *cal*, two genes with redundant functions in the triggering of floral development. The genes *tfl* and *lfy* inhibit each other, as do *ap-1* or *cal* and *tfl*. The apical expression of *tfl* obstructs the expression of floral genes and thus promotes indefinite growth (modified from Hempel et al., 2000).

appearance of the inflorescence, even though such a genetic anomaly does not appear to be sufficient (Purugganan et al., 2000).

2.2.6. Genes Downstream of Genetic Switching and the Evolution of Ovule

More than by its genetic characterization (A/B combination), a petal is defined phenotypically by its texture, colour, and other factors. Many genes are implicated in the development of floral organs once their identity is fixed: genes for pigment synthesis, genes for modification of the cell wall, etc. Studies have been conducted on the development of ovules and, even though our aim is not to describe this process in detail, we will mention that (1) the development of the ovule is triggered by factors D and genetic switching and (2) the genes thus activated can be classified into at least three groups (Schneitz et al., 1998):

— genes for development of the nucellus and megasporogenesis: *mac-1, msd, am-1, emd, fem, gfa, prl, ada, tya, hdd, Gf, lo-2, ig-1*;
— genes for development of the funiculus and chalaza: *hll, ant, mog, sub*;
— genes for development of the integument: *INO, bel-1, ucn, ats, sub, bag, sin-1, mog, tso-1, lug, lal, tsl, sup.*

Does the molecular biology of ovule development give us the clues to the history of Angiosperms? First of all, it seems quite legitimate to consider the ovule as a whole organ fabricated under the influence of homeotic genes of group D. The whorl of ovules (4*b*) is thus distinct from that of carpels (4*a*), but that is not to say that the factors of group C are not necessary to the development of ovules. It is in fact possible that *ag* will be required, in combination with *AGL11*, even though the results on this subject remain imprecise (Angenent and Colombo, 1996).

Remark: Genes of group D, i.e., *FBP7* and *FBP11* in petunia, are clearly homeotic genes: for example, mutants in which there is co-suppression of these genes develop carpels in place of ovules; the latter are fitted into carpels of whorl 4 and thus are slender, attain an elongated form, and become flat, which is why this mutant is called "spaghetti".

However, although it is true that in petunia the development of the ovule is independent of that of the carpels, the same does not apply in *Arabidopsis*, where the ovules form from the wall of the developing carpels. This poses a problem of evolutionary interpretation: does the formation of the carpels come from the closure of the telomes that carry the pre-ovules, or from the extension and closure of a "cupule" of foliar origin (in the first case the cupule would have formed the outer integument of the ovule)? The second case supposes an independent origin of the wall of the carpels and the ovule that is found in petunia. The first case suggests an order of development (first the carpel wall and second the ovules) that is found in *Arabidopsis*.

From some genes, the possibilities of the evolutionary pathway can be traced.

Acquisition of ovular integuments

In *ant* (*ain't tegumenta*) mutants of *Arabidopsis*, the integuments do not develop, so the nucellus follows an aberrant course of development: there is no embryo sac. This gene is expressed in all the primordia in the plant. In *bel-1* mutants, the inner integument does not develop until the outer integument develops abnormally, producing a structure of the carpelloid type. This observation suggests that outer and inner integuments develop independently and thus probably have different origins and that, given a form similar to that of the carpels, the cupule is the source of the outer integument. The *INO* (*inner no outer*) mutant has a complementary phenotype: no outer integument but development of the inner integument. Finally, the *sin-1* (*short integuments*) mutant has very short integuments, leading to an abnormal development of the embryo sac. Thus, two pathways that are *a priori* separate have probably been acquired over the course of evolution: one using a gene of the *INO* type allowing formation of the first integument of the ovule, the other using a gene of the *bel* type and forming the second integument. By means of a pathway that remains to be elucidated, these two integuments are controlled by common genes.

Ortho- / ana- or campylotropous transition

A key for this transition is represented by the gene *sup*, in which the integuments develop symmetrically, forming an ovule of tubular appearance recalling the orthotropous ovules. Thus, it is presently believed that a gene of this type would lead to asymmetry of integument development and thus the acquisition of the anatropous and campylotropous character.

2.3. FLORAL EVOLUTION THROUGH GENES?

2.3.1. Definition

Most ABC genes have a highly conserved part of around 180 base pairs called the MADS box (for MCM-1, *Agamous, Deficiens*, SRF). This part codes for the MADS domain of the corresponding protein, the role of which involves transcription factor activity. Unlike in animals, where the *hox* genes are organized into complexes, MADS genes are dispersed in the plant genome. The MADS domain is the determinant component of the protein-DNA interaction; its three-dimensional structure is a superhelix α (*coiled-coil structure*) that can be fitted in a small groove of DNA. The MADS box has a highly conserved sequence among living organisms, the consensus of which is *CC* $(A/T)_6$*GG*, called *CarG* (for *C-A-rich G*). In addition to the MADS domain, the MADS factors are composed of several other functional parts; these are parts I, K, and C. The MADS domain constitutes the "M" part. Thus, the MADS factors in the plants are said to have a *MIKC* structure.

I, which is immediately downstream of the MADS domain, is variable from one transcription factor to another. This part is implicated in the interaction with other transcription factors.

K, which seems specific to plants, is characterized by a succession of hydrophobic residues with more or less regular intervals, forming an amphipathic helix. It was suggested that this part allows dimerization of MADS factors (Theissen et al., 2000).

C is highly variable in sequence; its function is unknown.

Phylogenetic analysis of MADS genes of plants shows that it is a monophyletic group and that several lineages appear: those of genes of group A, group B, and group C. Genes of group D are contained in the "C line". Finally, as has been said above, there is an Im line.

2.3.2. Cormophytes and MADS Genes

a) Ferns

Even though MADS genes have been found in a moss and a Lycopod, those discoveries are yet to be confirmed. On the other hand, MADS genes have been identified and cloned in the ferns (Filicineae or Filicophytes). The studies conducted on the ferns cannot, however, be generalized because they apply only to the model plant *Ceratopsis richardii.* These are the *CRM* genes (*Ceratopsis richardii MADS*), which has a structure of the MIKC type (Theissen et al., 2000). Phylogenetic analysis of MADS genes suggests that Spermatophytes and ferns have a common ancestor, from which were formed the groups of MADS genes of modern plants by duplication and/or speciation. However, as there were already several lines of MADS genes in ferns, we cannot exclude *a priori* the possibility of a common ancestor with 2 or even 3 MADS genes. In any case, even if the MADS genes of Spermatophytes and Filicophytes are homologous, it is nevertheless possible that they may not be orthologous.

More precisely, studies of *in situ* hybridization have shown that in *Ceratopsis richardii* the MADS genes are expressed in the gametophyte as well as in the sporophyte, in the vegetative tissues in general, with the exception of three genes such as *CRM-1*, which are expressed in the sporangia, among other parts. This situation is very different from that of the Angiosperms, for which MADS genes are implicated only in the sporophytic development, with the exception of *AGL-17* in *Arabidopsis* (which is expressed in the pollen). Moreover, unlike in ferns, MADS genes are expressed generally in an organ-specific manner. Another character distinguishing MADS genes of ferns is the existence of alternate splicing for the overwhelming majority of them, which has been reported in only a single case in the Angiosperms. Some MADS genes of ferns, even though they have a structure of the MIKC type, also has STOP triplets or other anomalies. It is possible that the proteins formed are incomplete, in which case their function remains to be specified, or even that there are other processes at work such as RNA editing.

b) Gymnosperms

The MADS genes have also been studied in Gymnosperms and some homologues of *ag* and *pi/ap-3* of *Arabidopsis* have effectively been characterized. On the other hand, no homologous gene of *ap-1* (function A) has been found, which leads us to propose that the common ancestor of Gymnosperms and Angiosperms had a gene of the *ap-1* group that had been secondarily lost in the Gymnosperms, a hypothesis that is yet to be confirmed.

In phylogenetic terms, the MADS genes of the *pi* type of Gymnosperms are collected in a group related to the *pi* genes of Angiosperms. The transcription of these genes has been more particularly studied in *Gnetum gnemon*: while *GGM1* is expressed in the leaf as well as the flower, *GGM2*, similar to *ap-3* of *Arabidopsis*, is expressed in the male cones, and *GGM13* is expressed specifically in the nucellus, the integument, and the middle envelope of the ovule. It should be noted that in snapdragon, the gene *DEFH21*, a homologue of *GGM13*, is expressed in the inner integument (Becker et al., 2002).

The gymnosperm genes of group *ag* are better known. Five MADS genes of this group are orthologues of *C* genes of Angiosperms, and in *Epicea* (Pinaceae) at least two MADS genes have been cloned, *DAL-2* and *SAG-1*; it has been shown that *SAG-1* can complement *ag* mutants of *Arabidopsis* (Owens et al., 1998). In the Gnetales, the mRNA of *GGM3* is expressed in the male as well as female flowers.

Even though some aspects remain uncertain, it can be supposed that, as in Angiosperms, in the Conifers male cones are formed by means of the combination of homologues of B and C, and that female cones are constituted by the expression of C (which will also take on function D) alone. In the Gnetales, the intervention of a gene of group *ap-3/pi* in the determination of female structures introduces an additional complication. Nevertheless, the absence of genes of group *ap-1* allows us to rule out the hypothesis of homology between the ovular envelopes of *Gnetum* and the angiosperm perianth. Moreover, the expression of homologous genes (of group *ag*) *PrMADS2* and *3* in the ovule and the ovuliferous scale of *Pinus radiata*, as well as *GpMADS1* in the ovule of *Gnetum parviflorum*, suggests that the collar and ovule of Gnetaceae are respectively homologues of the bract and the combination of ovule and ovuliferous scale of Conifers (Hasebe, 1999).

Finally, it should be noted that ABC genes of Gymnosperms are expressed depending on orthologues of *lfy*. Two genes of this type have been identified in *Pinus radiata* and one of them, *nly* (*needly*), may complement the *lfy* mutants of *Arabidopsis* (Hasebe, 1999), which suggests that the function of *nly* in the Conifers is close to that observed in *Arabidopsis*. It is now supposed that the ancestor of Spermatophytes had two *lfy* genes, one of which was lost in the Angiosperms.

c) Angiosperms

Before tracing the principal traits relating to MADS genes of Angiosperms, we will give some information about the phylogeny that will serve as a basis for the subsequent discussion. This information will be brief, especially since the phylogeny will be explained in detail later in the book (Chapter 3).

Phylogeny of Angiosperms

These past few years, molecular phylogenies have shown that the Angiosperms constitute a monophyletic group and that they are organized in several lineages, some of which were predicted by systematists even before the advent of modern phylogeny. For example, there is the paraphyletic group of **non-monocotyledonous paleoherbs**, including notably the Nympheales and Piperales. Subsequently, there is a paraphyletic group called the **Magnoliidae complex**, including notably the Magnoliales. Then, there is the monophyletic group of Monocotyledons. Finally, there is the monophyletic group of Eudicotyledons, which begins with Ranunculales, includes Caryophyllales and Rosidae, and ends with Asterales.

Study of MADS genes in the Monocotyledons

The MADS genes and their expression have been the subject of very few studies in taxa other than model organisms such as *Arabidopsis*, petunia, or the snapdragon *Antirrhinum majus*, with the exception of maize (*Zea mays*, Poaceae). In Poaceae, the flowers have stamens and carpels but do not have petals or sepals. On the other hand, they have parts that look like bracts: the glumes, lemma, palea, and lodicules (Fig. 2.27). Morphological observations suggest that the glumes probably serve as bracts to the extent that they are arranged at the outermost part with respect to the other floral parts, and to the extent that they enclose some flowers in the form of a spikelet (small spike of 2 or 3 flowers). On the other hand, it is tentatively supposed that there is an equivalence between lemma/palea and sepals and between lodicules and petals. A decisive response was provided by the study of homologues of MADS genes of *Arabidopsis* in maize (Ambrose et al., 2000).

From the beginning of the 20th century, mutants of floral architecture were observed in maize (Fraser, 1933), such as *si-1* (*silky-1*), in which the stamens are replaced by female parts that look like bracts, and the lodicules by parts

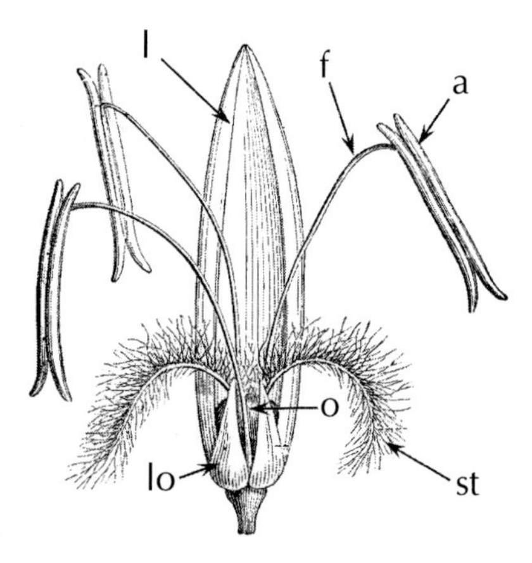

Fig. 2.27: Structure of a flower of Gramineae in which the palea has been removed. l, lemma; f, filament; a, anther; o, ovary; st, stigma; lo, lodicule (modified from Duchartre, 1885).

resembling palea. More precisely, the stamens have a marked vascular network and stigmatal extensions. These flowers do not open because they are blocked by the development of palea and lemma in place of lodicules. Moreover, the mutants generally have several pistils: one central and three others developing in the place normally occupied by the stamens. These last are nevertheless sterile. The mutation *si-1* thus recalls the mutants for function B in *Arabidopsis*. The correspondent gene, *si-1*, was cloned and presents a significant identity with *ap-3* of *Arabidopsis* or even *def* of *Antirrhinum*. This gene has a MADS gene structure, and the protein that it codes has a structure of the MIKC type. *In situ* hybridizations have confirmed that *si-1* is expressed clearly in the primordia of stamens and lodicules.

These studies on the *si-1* gene were part of the latest data that completed the model of genetic switching in the Gramineae, which resembles the genetic switching of *Arabidopsis*: there is a similar switching with three functions, in which C is represented by Zag-1 and Zmm-2, B by Si-1, and A by unknown factors (Fig. 2.28). It was remarked that C comprises two factors, while Ag has

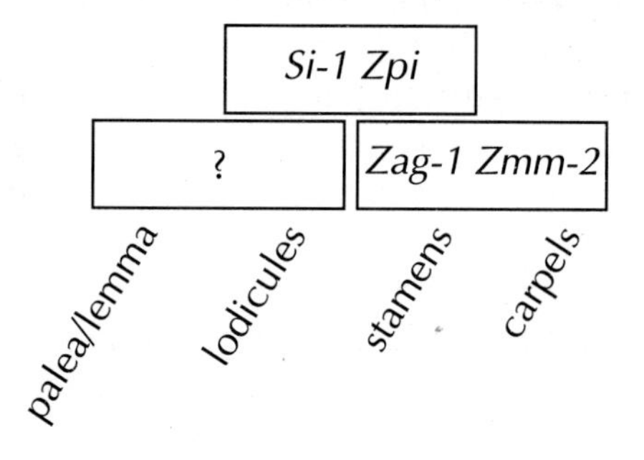

Fig. 2.28: Genetic determination of the setting in of whorls in maize (see text for details)

only this function in *Arabidopsis*, and that the function B comprises only a single factor, while *zpi* is identified lastly as a gene having significant identity with *pi*. So far there are no practical data on function A in maize, nor on a possible D function. However, a homologue of *ap-1* of *Arabidopsis* was discovered in another Gramineae, *Lolium temulentum*: *LtMADS2* (Gocal et al., 2001).

Since then, even though there are some aspects to be studied further, it seems clear that the genetic switching of type "ABC" is conserved during the course of the evolution of Angiosperms, since it is found in Monocotyledons, in which function A specifies the paleae, the A-B combination the lodicules, the B-C combination the stamens, and C the gynoecium. This demonstrates that there is a sepal-palea/lemma equivalence and a petal-lodicule equivalence.

MADS genes and phylogeny of Angiosperms

From the time the genetic switching of type ABC was found to be undoubtedly highly conserved in Angiosperms, as has been suggested, and considering the floral architecture of the major groups (see above), it was possible to formulate hypotheses concerning the evolution of whorls and its parallel in terms of expression of MADS genes. We will later rely on the model of gene duplication-speciation, HYPAG (*hypothesis of parallel architecture—MADS genes*). To approach this, it is advisable to start with the floral structure of different groups.

MADS Genes in Other Angiosperms

Apart from *Arabidopsis*, snapdragon (*Antirrhinum*), and petunia, MADS genes have been specifically identified in tomato, asparagus, and more recently in eucalyptus. The genes *egm1*, *egm2*, and *egm3* of *E. grandis* are expressed only in the flowers, *egm3* being expressed also in the receptacle. In particular, *egm2*, very similar to *globosa* of snapdragon, is expressed in the petals and stamens (Southerton et al., 1998). The other genes present less clear profiles of expression. Thus, the genetic switching system also seems to be present in the Myrtales. Comparable genes (*CUM* genes) have been studied in cucumber (Cucurbitaceae). The profiles of expression are also similar to those observed in *Arabidopsis*. However, there are two C genes in cucumber: *CUM-1* and *CUM-10* (Kater et al., 1998). Finally, there are MADS genes expressed most often in the entire floral primordium and finally in all the organs of the flower, the functions of which remain to be specified: this type of gene has been indicated in Orchidaceae (Yu and Goh, 2000) as well as in Rosaceae (Sung et al., 1999).

— Piperales: A prototype of this group is *Peperomia* sp., in which the flowers are bisexual but very simple because, apart from the bract, there are two reduced stamens (in *P. glabella* for example) and a pistil, i.e., they have no perianth. The floral formula can be written simply as follows: **, -0-, 1-10,*

1-4, drupe. The absence of the perianth suggests that there is no function A in these plants.

— Nympheales: A well-known flower of this group is the water lily, an actinomorphic flower, presenting many tepals, *n* stamens, *n* joined carpels, and very often parts intermediate between stamens and tepals (Fig. 2.29).

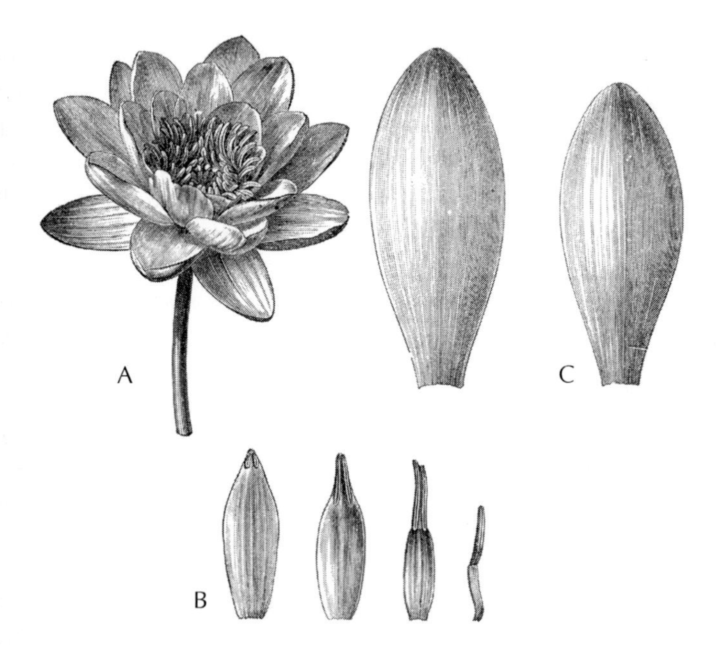

Fig. 2.29: External view of white water lily (A), with different floral parts intermediate between stamen and tepal (B, C). The floral parts go from outer to inner from the left to the right (modified from Duchartre, 1885).

Remark: The progressive transformation of tepals into stamens as we go towards the interior of the flower was used by Goethe to argue in favour of his **metamorphosis theory**. His idea of the metamorphosis of tepals into stamens is revived today in the form of genetic swtiching, which specifies the identity of organs from cell masses having *a priori* the same potentialities (as the homeotic mutants attest). An example of floral "metamorphosis" well known to horticulturists is that of **double flowers**, in which, most often, it is the stamens that are converted into petals, giving an impression of "doubling" of the perianth thickness.

— Liliales: A typical family of this group is Liliaceae, including in particular the lily, a plant of type 3 having 3 petaloid sepals and 3 petals (or 6 tepals), 6 stamens, and 3 united carpels. In this case, the external tepals are believed to be sepals that are undoubtedly modified secondarily into petals. As in the Nympheales, we are led to suppose that over a large domain,

A and B are expressed simultaneously, and that there is no whorl where there is only function A.

- Poales: Gramineae constitute a good example, which has been detailed above in the case of maize. There is reason to conclude that the sepal-petal-stamen-carpel organization in 4 whorls very likely exists in these plants but with great modification.
- Eudicotyledons: *Arabidopsis thaliana* (Brassicaceae) or even the snapdragon *Antirrhinum majus* (Plantaginaceae) are typical examples.

HYPAG (hypothesis about the phylogeny of genes determining floral architecture) can be summarized as in Fig. 2.30, a highly simplified version of real situation. The Gymnosperm-Angiosperm divergence was likely to be accompanied by the acquisition of a function A. The duplication of a gene of the *agamous* type would have generated a gene with a function C and a function D, two functions that are apparently served by the same gene in the Gymnosperms. Within the context of this figure, in the Piperales, function A is lost. The absence of the expression of function A **only** in the first whorl or whorls leads to a perianth constituted solely of tepals.

Finally, it must be noted that recent dissection of MADS genes in the Eudicotyledons indicates that it is only from "central" Eudicotyledons (from Vitales to Asteraceae, see Fig. 3.1) that the expression of B genes is similar to that observed in *Arabidopsis*. In the Palaeoherbaceae, the Magnoliidae, and the basal Eudicotyledons, genes of group *ap-3* are expressed also in the carpels. These genes are distinguished phylogenetically as well, forming the group

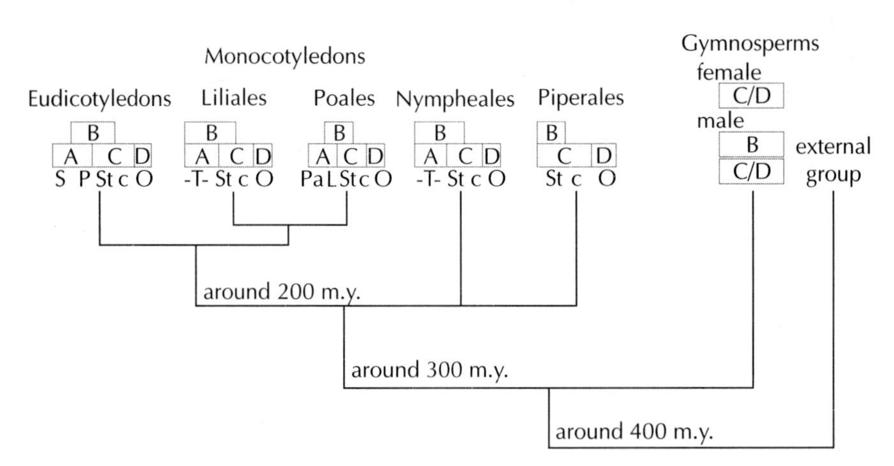

Fig. 2.30: Hypothesis about the phylogeny of genes determining floral architecture (HYPAG, see text). Associated genetic functions are indicated with their initials A, B, C, D, and C/D (a common function of Gymnosperms). S, sepal; P, petal; St, stamen; Pa, palea; L, lodicule; -T-, tepal; c, carpel; O, ovules. The chronological data are indicated in millions of years (m.y.) (modified from Theissen et al., 2000).

palaeo ap-3 or B sister (Kramer and Irish, 2000). It is supposed that there was a genic duplication from an ancestral gene *palaeo ap-3* giving a line *eu ap-3* (such as *ap-3* of *Arabidopsis*) and a line of type *TM6*.

2.3.3. Genes and Floral Type

Mutants of *Arabidopsis thaliana* present an abnormal number of floral organs or whorls. This is the case with *clavata* (*clv*), which has a wide floral meristem and thus more whorls. In some cases, the ectopic mutants *35S:ap-3* have one or more supplementary whorls of stamens. In the mutant *sup* (*superman*), supernumerary stamens are found in whorls 3 and 4. Thus, it is possible that the number of stamens is fixed in the flower by the antagonism between factors Sup and Ap-3/Pi, and particularly by means of the inhibition of *ap-3* and *pi* by Sup in whorl 4 (Howell, 1998). However, experiments of *in situ* hybridization have shown that *sup* is expressed at the frontier between whorls 3 and 4. Moreover, the presence of supernumerary stamens suggests a number of cell divisions that is greater than normal at this level, which suggests that *sup* will repress cell division in whorl 3 and activate it in whorl 4. The validity of this model proposed by Sakai (1995) remains to be studied, however. Another gene, *fon-1* (*floral organ number 1*), seems to control the number of floral organs of whorls 3 and 4, because the mutant *fon-1* has more stamens and carpels than the wild type.

The mutant *perianthia* (*pan*) of *Arabidopsis* is much more interesting, in that it has flowers of type 5 that are perfectly formed (floral formula **, 5, 5, 5,* ②), and not of type 4 like the wild type (**, 4, 4, 4 + 2,* ②). In the mutant *pan*, the petals alternate with the sepals, and the stamens are episepalous. The bract faces a petal. The flower of the mutant thus presents a typical organization of type 5. The details of the mode of action of the *pan* gene remain to be determined. Nonetheless, the acquisition of this gene (and possibly of other genes associated with it) is responsible for the original structure [*4, 4, 4 + 2,* ②] of Brassicaceae. Besides, the existence of this gene, as well as that of the ABCD system that is found also in other species, probably puts an end to many theories that see the origin of the flower of Brassicaceae in an inflorescence that was highly modified and condensed (for a review, see Vallade, 1999).

3

Evolution of Floral Structure

To discuss floral evolution in Angiosperms we must go back and study the history of these organisms, which requires us to look at **phylogeny**. This is why the discussion begins with a look at the phylogenetic tree (consensus) of Angiosperms, in order to draw out the most important concepts for the interpretation of floral structures. The reverse approach (reconstructing a phylogenetic tree from the floral characters) would perhaps be more pedagogic; however, it demands an extensive knowledge of botanical groups. In addition, the phylogenetic trees used at present have also been established by molecular data that are not intuitive.

3.1. PHYLOGENY OF ANGIOSPERMS

3.1.1. Principles of Constructing a Phylogenetic Tree

a) Definitions

To put it very simply, there are two types of phylogenetic constructions (or **trees**): **cladograms** and **phenograms**. The two types are named after the method by which these trees are obtained. Cladograms use binary characters of the yes/no or 0/1 type, for example the persistence of a characteristic (e.g., actinomorphy) and its loss (in this case zygomorphy), or ternary and more generally **discrete** characters. At each bifurcation of the tree, a characteristic of the taxon takes one value on one branch and another on the second branch. Phenograms, on the other hand, are constructed by means of calculations of distances such as genetic or morphological distances. The length of branches is generally proportionate to the distance calculated. A phenogram is constructed using a **matrix of distances** between the objects of phylogenetic study. In theory, it is possible to construct a phenogram "directly" from one relative to the next calculating as one goes along the matrices between the barycentres (nodes) and the remaining objects. However, algorithms can be used to draw phenograms using a computer. The same applies to cladograms. In general, the trees are constructed using the principle of **parsimony**, which stipulates that the greatest possible economy is to be observed in the number of changes of state of characters (in particular, careful attention must be given to supposed **reversions** of characters—see below).

b) Traps to avoid

The very brief information given below can still be used to anticipate some problems that one should beware of.

— Phylogenetic trees are sensitive to **initial conditions**, i.e., to the objects used to construct them. The addition of even a single object to a set of objects organized into a phylogenetic tree may significantly modify that tree. Obviously, the modification of characters used also changes the tree under consideration. We must therefore pay close attention to the **strength** of phylogenetic trees.

— The node separating two branches of a phenogram **does not represent** the common ancestor of the two objects considered. It is rather a medium (barycentre). On the other hand, the existence of a node **suggests** that the two objects probably had a common ancestor, which is **presently unknown**.

— Often, terms such as "evolved" and "primitive" are used. A species that branches out near the base of the phylogenetic tree **is not** primitive. Since it is still observed today, it has undergone the effects of evolutionary processes, like all other species. For example, it is not correct to say that the maritime pine is less evolved than the daisy. The pine has certainly evolved since its appearance, the pine today being different from that of earlier times, at least genetically. Moreover, the existence of a very long branch in a phylogenetic tree, as is the case with species that branch out at the base, does not necessarily indicate that the evolution has also been very long and thus that it is a species that emerged very early in the course of evolution. The interpretation of trees on the basis of distances starts from the following principle: if the distance (particularly the genetic distance) is very long, then the species emerged early. But this does not apply if the speed of evolution of the branch considered is high. This phenomenon, **the attraction of large branches**, has been the source of unfortunate confusion. Eventually, the species emerging at the base of the tree should be called not "primitive" but **conservative** (which is more exact from a phylogenetic point of view).

— Phylogenies are based on the underlying idea of **homology**. Two objects are homologues if they have a common origin (if there is a common ancestor). Conversely, two objects that are very similar (two species having a trait in common) do not necessarily have a common ancestor. For example, on observing a reversion from zygomorphy towards actinomorphy in species A, we may place A in a group that has never known a transition towards zygomorphy. This involuntary error is an example of **homoplasy**.

c) Parsimony, weight, reliability

The construction of a phylogenetic tree is always based on a certain number of hypotheses, which mostly have to do with the **probability** of change of state of a character. Let us suppose, for example, a discrete character having more than two possible states. How do we take into consideration the transition from one state to another? Does the transition follow a **given order** or not? The existence of an imposed order reduces the field of possibilities of successive transitions (i.e., changes of state) and is thus responsible for a certain "economy", or **parsimony**, in the words of **Wagner**. When such an order does not exist, we refer to the **Fitch procedure**.

Moreover, the state of a character may change and then return to its initial state. This is called **reversion**. Another particular case concerns the independent acquisition of common characters by two separate **phyla**, which is called **parallelism**. These two processes, reversion and parallelism, must be attributed probabilities during the construction of a phylogenetic tree. The **Dollo method** supposes reversions to be easy (the mutation or deletion of a single gene may lead to the "loss" of a function) and parallelisms to be improbable (low probability that two comparable gene sequences would appear independently). The definition of "loss" or "gain" in the Dollo method supposes an original state, i.e., that the trees established in this way are necessarily **rooted** (trees without roots are called **networks**).

Sometimes, during cladistic analyses, the characters are **weighted**. The problem thus lies in the attribution of weights to different characters. One method, called successive weighting, consists in making a preliminary phylogenetic tree by giving an identical weight to all the characters. On the shortest trees obtained, we mark the characters presenting the least homoplasy. Such characters are attributed the greatest weight. There are other methods using genetic data (for example, it is known that transitions and passages do not have the same frequency in nature). However, phylogenetic trees are often constructed **without weight**. Moreover, it is quite frequently supposed that the change of state of a character **does not affect** the probability of change of state of another character, which is undoubtedly not the case in reality.

Finally, what credit can be accorded to a phylogenetic tree? A first glance at a tree must especially look at the characters susceptible of generating homoplasy. This would allow us to rapidly calculate a **confidence index**, which is equal to the minimal number of evolutionary changes (especially genetic ones) divided by that actually observed in the tree. The confidence index thus varies from 0 (doubtful) to 1 (confident). A problem with this index is that some characters are not informative (e.g., a single derived character, i.e., autapomorphy—see below). These characters will somewhat "uselessly" increase the confidence index. Also, other indexes are used, such as the index of retention r:

$$r = \frac{M - L}{M - m}$$

where L is the real length of the tree, M is its maximum length, and m is its minimum length. L refers to the number of real changes in the tree and m to the minimum number of changes. M is the number of changes that would be observed if all the taxa having a derived character were independent.

There are statistical methods that can be used to give information about the confidence that can be placed in phylogenetic trees (*bootstrap*, etc.). Moreover, one way to know whether a tree is reliable is to compare it with very different data (e.g., morphology and nucleotide sequences of chloroplast DNA).

Finally, it is possible to "prune" a phylogenetic tree. Indeed, phylogenetic analyses generally provide several possible trees. Some rules are available by which to choose the most suitable tree:

- the most simple: this is the **rule of parsimony**;
- the most likely: the **maximum probability** method is used;
- the most economical in terms of distance: the total distance is minimized (several mathematical methods can be used).

However, several trees can still concur. In such a case, a tree "summarizing" all of them can be drawn, called a **consensus** tree, concentrating the common information and calculating the uncertainties. The **strict** consensus trees contain only the **monophyletic** groups common to the possible trees.

3.1.2. Descriptive Elements of Phylogenetic Trees

a) Synapomorphy, symplesiomorphy and autapomorphy

A special terminology is used to describe phylogenetic trees and especially to describe the characters of species. A trait is said to be **synapomorphic** when it represents the state of a character that is derived and common to an entire group. The phylogenetic significance of a group is based mostly on the existence of a synapomorphy. A **symplesiomorphy** is a character conserved in the ancestral state. An **autapomorphy** is a character shared by a single group, without a correspondence in other groups. That means that an autapomorphy is of no use to the phylogeneticist. For example, the hyperstigmas of

Monimiaceae are peculiar to that family and thus give us no useful information in relating the Monimiaceae to other families.

b) Paraphyly, monophyly and polyphyly

Once the tree is constructed, it still must describe the groups obtained. A "group" is a set of *k* species. A group of species sharing a single common ancestor is said to be **monophyletic**. A group of species sharing several common ancestors is said to be **polyphyletic**. In the two cases, the ancestor(s) give rise to exactly *k* species. If there are one or several species of the group that are not descended from those ancestors, the group is **paraphyletic**. The same applies if the ancestors have given rise to other species that do not belong to the group considered. It is easy to see that the legitimacy of a group (and its **name**) is clear only in the case of monophyletic groups, since the grouping takes on a historic dimension. It is especially remarkable that the Bryophytes, Pteridophytes, or even Pteridosperms are paraphyletic groups.

3.1.3. The Phylogenetic Tree of Angiosperms

a) A brief history

Early studies on the phylogeny of Angiosperms were numerous and will not be discussed in detail here. The first attempt at classification appeared during the Renaissance (notably by Caesalpino, 1519-1603, and other naturalists). The existence of botanical families was first suggested by Magnol in 1689, and their identification was based on morphological characters of the entire plant, or on the presence of an undefined "perceptible affinity" understood mainly by intuition. The concepts of **species** and **genera** were established much later. Tournefort (1656-1708) proposed that plants having common traits be grouped in genera of order 1 and, within each of these, plants having common characters (not shared by other plants of the genus of order 1) be grouped within genera of order 2. According to Linnaeus (1707-1778), genera were "natural" entities, presenting a certain number of precise characters: *characterem non constituere genus, sed genus characterem*, i.e., a genus exists in nature, independently of the characters used to describe it. Linnaeus invented **binomial nomenclature** to name the species, which he also believed to be "natural" (it was only in the 20th century that it was acknowledged that genera and families were more "artificial" than species). Linnaeus used only species and genera to establish a classification of plants based essentially on the flowers and fruits, taking into consideration the number of stamens and the arrangement and composition of the gynoecium (number of stigmas, styles). He constructed 67 families and was unfortunately left with a "remainder" of a few unclassified genera. The studies of the great botanist Jussieu (1748-1836) led to the establishment of families the characters of which remained sometimes fluid. Jussieu considered that species constituted an evolutionary continuum, and that it was thus difficult to establish the limits of a particular group within that continuum.

Even though the establishment of botanical groups was not without difficulty, scientists also sought to organize the various groups by means of **probable filiations**. They felt the need to find characters for this purpose, and Adanson (1727-1806) was the first to emphasize the problem of the **weight** of such characters, suggesting that no character is inherently essential in relation to others, in separating botanical groups. Jussieu did not see the problem in the same light, arguing that it was possible, gradually, with **similarities** between species, to construct more and more smaller groups, showing by this that characters did not all have the same value since they were applied at different levels. In 1778, Lamarck (1744-1829) proposed a sort of tree that would give a numerical weight to characters. Filiations between existing families were proposed, especially in what were called the "cactus" diagrams of Bessey (1845-1915). Dahlgren published "parts" of phylogenetic trees (dahlgrenograms), representing clusters of botanical groups. However, the relationships between these groups remained imprecise.

Around 1940, Glimour (1906-1986) remarked that the characters taken into account to establish botanical groups were overall the same as the those used to establish putative filiations, which demonstrated a tautological process. It was therefore necessary to find common characters independent of those used to identify or classify. The principles ultimately used in phylogeny were established by Hennig (1950) and Wagner (1969) (see above).

b) Present-day phylogeny

General points

We first present a phylogenetic tree of major groups of Angiosperms based on the morphological characters and the sequences of genes *rbcL*, *atpB*, and ribosomal genes *18S* (Soltis et al., 1998). Speaking broadly, it can be said that there is a **division of Dicotyledons** (Fig. 3.1). That is to say, the Dicotyledons

---

Fig. 3.1: Phylogenetic trees showing the major groups of Angiosperms (A), Monocotyledons (B), Eudicotyledons (C), and Asteridae (D). Tree B was obtained by alignment of sequences of the gene *rbcL*. Trees A, C, and D were obtained by means of combined characters derived from the morphology, sequences of genes *rbcL*, *atpB*, and ribosomal genes *18S*. (A) The exact place of orders Illiciales and Ceratophyllales is uncertain, and it has been proposed that they constitute, along with Nympheales and Austrobaileyaceae, a sister group of the set of other Angiosperms. In an attempt to simplify, these orders are here presented in the form of a "rake". The Amborellaceae are not integrated with this phylogeny and their place is discussed later (see Fig. 3.2). (B) The Arales (group comprising Araceae) are considered included in the Alismatales *s.l.* The order Typhales also includes the Sparganiaceae (not represented in the figure). (C) The Rosidae are sometimes grouped into Euasteridae I (Garryales, Solanales, Gentianales, and Lamiales) and Euasteridae II (Aquifoliales, Apiales, Dipsacales, and Asterales), two monophyletic groups. (A, C, D modified from Judd et al., 1999 and B from Bremer et al., 2000.)

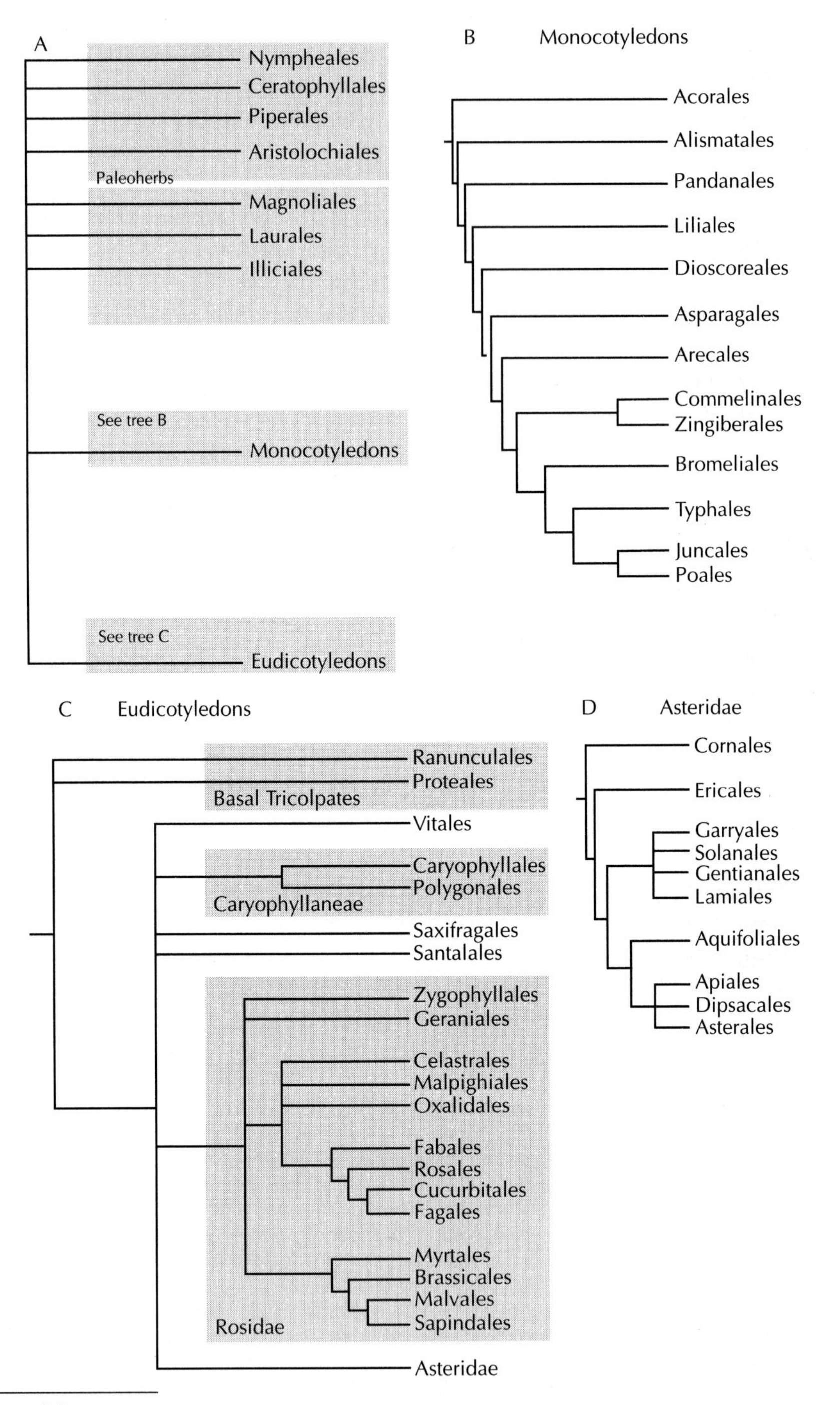
A
Nympheales
Ceratophyllales
Piperales
Aristolochiales
Paleoherbs
Magnoliales
Laurales
Illiciales
See tree B
Monocotyledons
See tree C
Eudicotyledons
B Monocotyledons
Acorales
Alismatales
Pandanales
Liliales
Dioscoreales
Asparagales
Arecales
Commelinales
Zingiberales
Bromeliales
Typhales
Juncales
Poales
C Eudicotyledons
Ranunculales
Proteales
Basal Tricolpates
Vitales
Caryophyllales
Polygonales
Caryophyllaneae
Saxifragales
Santalales
Zygophyllales
Geraniales
Celastrales
Malpighiales
Oxalidales
Fabales
Rosales
Cucurbitales
Fagales
Myrtales
Brassicales
Malvales
Sapindales
Rosidae
Asteridae
D Asteridae
Cornales
Ericales
Garryales
Solanales
Gentianales
Lamiales
Aquifoliales
Apiales
Dipsacales
Asterales

Figure 3.1

are a paraphyletic group. In fact, it is observed that they are now organized into at least three groups: Magnoliidae, "paleoherbs", and Eudicotyledons. The Monocotyledons, on the other hand, are monophyletic.

The Magnoliales and Nympheales constitute two groups that branch out near the base of the tree. This does not mean that these plants are the ancestors of Angiosperms but only that they are rather "conservative". Moreover, the origin of Angiosperms is still unresolved, since, according to different phylogenies, the Gnetales emerge as a sister phylum of either Angiosperms or Conifers (see also the Introduction). The hypothesis of ancestral Magnoliales emerges from very simple observations, which place them in a position straddling Gymnosperms and Angiosperms. Notably, these plants have only tracheids with scalariform pits. Besides, the pollen grains have a wall with a granular structure (see below). Phylogenetic studies have added to the phylogeny in showing that the Amborellaceae constitute a family that branches out at the base of the dendrogram of Angiosperms and can be associated with the Nympheales (Soltis et al., 1999). The rooting of the dendrogram of Angiosperms indeed poses a problem, since it is seen that the Amborellaceae either branch out alone at the base or are associated with the Nympheales.

One study, using methods developed in the 1990s, seems to have an answer to this problem in simultaneously using genetic data from several cellular compartments (Barkman et al., 2000). We will give some information about these methods since they will undoubtedly be widely used in future phylogenetic studies. After determination of optimal outgroups selected by OOA or optimal outgroup analysis (Lyons-Weiler et al., 1998), the phylogenetic "noise" is reduced (the noise, which reduces the clarity of the phylogenetic signal of the tree, is caused by characters subject to problems: saturations, errors, attraction of major branches, etc.). Concretely, these methods use a measure of the phylogenetic "signal", the RAS (for relative apparent synapomorphy score). This score gives the number of times two taxa *i* and *j* share a state of a character, which is not presented by other taxa (Lyons-Weiler et al., 1996). Thus:

$$\text{RAS}_{i,j} = \Sigma_{k \in A} \, \Sigma_{c=1 \text{ to } n} r$$

where A is the number of taxa considered in the data and c is the character number. If $E_A(c)$ is the state of character c in taxon A, we have $r = 1$ if and only if $E_i(c) = E_j(c) \neq E_A(c)$ for every A different from i and j. This type of calculation makes it possible particularly to eliminate problems of convergence (the emergence of "false synapomorphies"), since the RAS quantifies the uniqueness of the characters taken into account to bring two taxa together. Two similar taxa have a high RAS, while two more distant taxa have a low RAS, even if there are one or more convergent states of characters. To put it simply, the reduction of "noise" means the elimination of characters introducing a reduction of the overall RAS, just as the OOA consists of choosing the outgroup giving an optimal overall RAS.

The two trees in concurrence (Amborellaceae alone at the base versus Amborellaceae + Nympheales, Fig. 3.2) have been tested by the bootstrap method and the problems indicated above have been reduced beforehand (reduction of noise, OOA). It seems clear that there is *greater* confidence about the rooting *associating Amborellaceae and Nympheales* than about that isolating the Amborellaceae. As emphasized by Barkman et al. (2000), from an evolutionary point of view, these considerations on the rooting have modest repercussions: the basal Amborellaceae would signify that the common ancestor of Angiosperms had unisexual flowers, while both the possibilities remain open in the "Amborellaceae + Nympheales" hypothesis (the Nympheaceae have hermaphrodite flowers). Moreover, *Amborella* has no vessels, which is not the case with Nympheales, and thus this new rooting suggests that the ancestor of Angiosperms certainly had vessels, and that *Amborella* lost them secondarily.

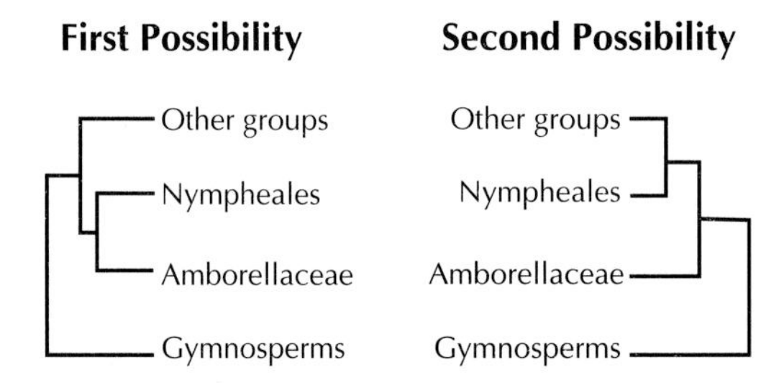

Fig. 3.2: Two possible modes of Angiosperm rooting combining the Nympheaceae and Amborellaceae (first possibility) or not combining them (second possibility).

The importance of pollen grains

The Eudicotyledons coincide with the **tricolpate** Angiosperms (the pollen grain of which has three **colpate apertures**, i.e., three pores, each set in a groove). In fact, several pollen types can be distinguished on the bases of **apertures** (Table 3.1, Fig. 3.3).

Table 3.1. Various pollen types

Type	Description	Example
Monosulcate	A single groove	Magnoliaceae
Monoporate	A single pore	Poaceae
Trisulcate	Three meridian grooves	Ranunculaceae
Tricolpate	Three pores each set in a groove	Many families
Triporate	Three pores	Cucurbitaceae
Polysulcate	More than three grooves	Cactaceae
Polyporate	More than three pores	Convolvulaceae

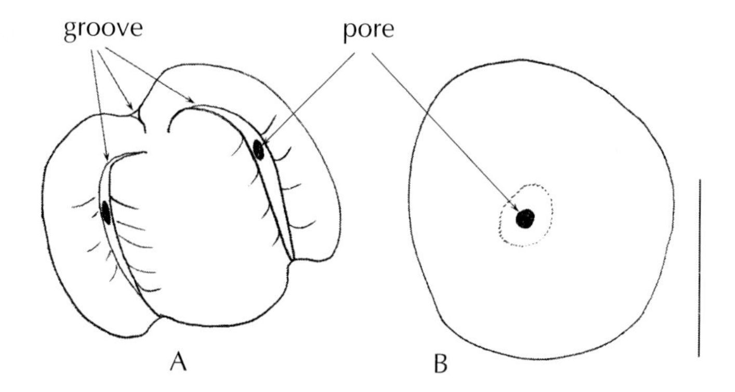

Fig. 3.3: Diagram of tricolpate pollen grains (A) and monoporate pollen grains (B). The bar is around 20 μm.

It has been supposed that evolution proceeds from monosulcate to tricolpate (Le Thomas, 1999), the Magnoliales having monosulcate pollen and the "evolved" Dicotyledons generally having tricolpate pollen. The Monocotyledons thus would have pollen with a derived monoporate structure.[1] The phylogeny shows that the trisulcate character is a synapomorphy of Eudicotyledons, the monoporate character probably being a synapomorphy of Monocotyledons. The plesiomorphic (ancestral) state of Angiosperms was probably a **monosulcate** pollen. Finally, there are Angiosperms in which the pollen have no apertures, as is typical of Lauraceae and Monimiaceae, two sister families of Laurales.

The Monocotyledons

The Monocotyledons constitute a monophyletic group, characterized especially by leaves with generally parallel venation, an embryo with a single cotyledon, and an adventitious root system (Judd et al., 1999). However, these characters considered independently are not synapomorphies of Monocotyledons (for example, some Nympheaceae have an adventitious fasciculate root system). The monophyly of Monocotyledons is visible on trees constructed on the basis of genetic as well as morphological data.

Paleoherbs and Magnoliidae

The paleoherbs constitute a paraphyletic group made up of different lineages, some elements of which we will describe, because these plants are often poorly known.

---

[1]Moreover, the pollen grain wall has two types of structure under transmission electron microscope: *granular* or *columnar*. The Gymnosperms have a granular wall, as do the Magnoliidae, excepting Winteraceae.

Nympheales: This order practically contains only one family, the Nympheaceae.[2] These are herbaceous plants with rhizomes, having an aerenchyma and often laticiferous ducts. The trichomes generally produce mucilage. The flowers have tepals (see Chapter 2, Fig. 2.29). The lotus, and more generally the Nelumbonaceae, even though they look similar to Nympheaceae, are classified among the Eudicotyledons (Proteales).

Ceratophyllales: This order comprises only one monophyletic family, the Ceratophyllaceae, a group of aquatic plants generally without functional roots. The leaves are whorled and dissected in a dichotomic manner. They have unisexual flowers (monoecious) with the following formula: male*, -7 to 8-,10-8, 0 and female *, -7 to 8-, 0, $\underline{1}$, achene.

Piperales: This order, characterized by several circles of cribro-vascular bundles, vessel elements with simple pits, and inflorescences in dense spikes, essentially comprises two families: Piperaceae (pepper, peperomia) and Saururaceae.

Aristolochiales: This order is composed of Aristolochiaceae and Lactoridaceae. The Aristolochiaceae produce aristolochic acids and have an inferior ovary (synapomorphies of the family).

The Magnoliidae comprise three orders: Magnoliales, Laurales, and Illiciales. The most important families are, respectively, Magnoliaceae, Lauraceae, and finally the Illiciaceae along with Winteraceae. Some aspects of these families are summarized later in the chapter.

The Eudicotyledons

The Eudicotyledons are organized into "basal tricolpates", a paraphyletic group comprising the Caryophyllales, and "central tricolpates" including notably the Rosidae and Asteridae (referred to as Sympetales). The Asteridae comprise the Cornales, Ericales, and Euasteridae I and II. Euasteridae are named for the star-like shape of their flowers. They are centred respectively around the Solanales-Lamiales, Gentianales, and Apiales-Asterales.

The monophyly of Eudicotyledons is based on analyses of the *rbcL* and *atpB* gene sequences (Soltis et al., 1998).

3.2. EVOLUTION OF FLORAL TRAITS LINKED TO THE GYNOECIUM

3.2.1. From "Incomplete" Angiospermy to Angiocarpy

The hermetic sealing of carpels, isolating the ovule or ovules from the external environment, is a characteristic of Angiosperms. However, this sealing could

---

[2]The Nympheaceae include the former families Barclayaceae and Cabombaceae, which are now ranked as sub-families. The Nympheaceae thus have five sub-families.

be limited to a very tight adhesion of the flanks of the carpels, as in the case of Magnolia (Fig. 3.4). This is why the Magnolia has often been considered primitive in relation to other Angiosperms. An imperfect, even incomplete, sealing has been supposed in some Winteraceae (this point is discussed in Endress, 1994), ranking this family much earlier than the Magnoliales in an evolutionary scale. However, the phylogeny does not show that these Magnoliidae (i.e., Winteraceae) branch out from the base of the dendrogram at a level lower than that of other Angiosperms. It is thus possible that the ancestor of Angiosperms had a more "perfect" suture than that of the Winteraceae, which still does not suggest that certain characters so far supposed primitive can now be considered derived. The Resedaceae also have a carpellary suture that is still described as imperfect; for this family, which emerges among the Brassicales, this is a **derived** character.

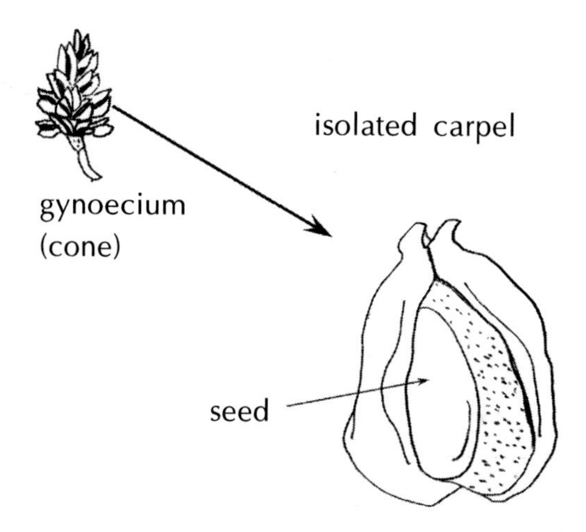

Fig. 3.4: Diagram of a carpel of magnolia seen from the outside. At maturity, the two lips of the dehiscent carpel separate, leaving the seed visible.

The carpels may be found enveloped in a supplementary covering that develops from the receptacle around the gynoecium. The ovary that ultimately results is **inferior**. This supplementary covering or **hypanthium** (or **torus**), forming a sort of floral cup, exists in the Rosaceae in particular, where it is clearly visible.

There are several possible configurations of the hypanthium (Fig. 3.5):

- In a **perigynous** flower, the hypanthium and the walls of the ovary (carpels) are not united, so that the ovary remains **superior**.
- In an **epigynous** flower, the ovary is adnate to the hypanthium and the ovary is **inferior**.

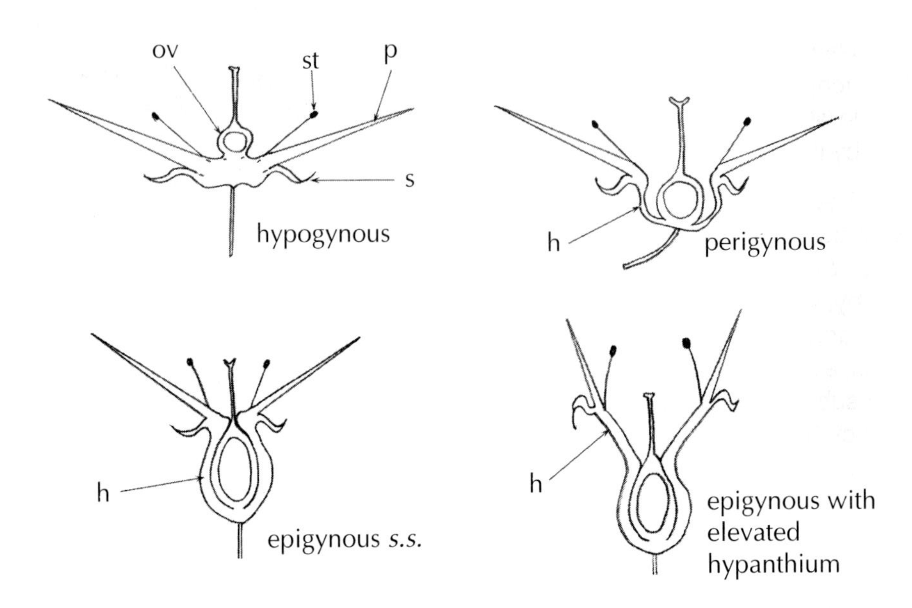

Fig. 3.5: The different configurations of the hypanthium (see text for explanations). h, hypanthium; p, petals; s, sepals; st, stamens, ov, ovary.

— In an **epigynous** flower with an elevated hypanthium, the hypanthium is highly developed, beyond the ovary, bearing stamens, petals and sepals further upward. The ovary is more clearly inferior.

Remark: There are presently two theories explaining the formation of epigyny: one is referred to as **receptacular**, the other **appendicular**. According to the first theory, the hypanthium originates in an expansion of the receptacle. According to the second, it results from the basal fusion of several organs, continuing their zones of attachment upward. The morphogenetic studies do not lead to a general rule: there are some families in which the formation of the hypanthium is receptacular (Rosaceae) and others in which it is appendicular (*Downingia*, Campanulaceae; Raven et al., 2001).

The development of a hypanthium leads to the enclosure of the ovary in a supplementary covering, whence the term **angiocarpy**. In phylogenetic terms, the presence of a hypanthium is a **synapomorphy** of Rosales (Judd et al., 1999). This is the case with Rosaceae, Rhamnaceae, and Ulmaceae. The hypanthium has been secondarily lost in the Celtidaceae, Moraceae, Cecropiaceae, and Urticaceae.

Remark: In some groups, the inferior ovary has been supposed to be an adaptive advantage, as with **ornithophilous** Zingiberales or Asparagales, in which the inferior ovaries are protected against possible injuries caused by the pollinating birds.

The Rosaceae are an interesting family from this point of view because they have different conformations of the hypanthium (Fig. 3.6). In the Spiroideae (e.g., *Spiraea, Sorbaria*), the hypanthium is not easy to see and the flowers are **hypogynous**. In the Maloideae (e.g., *Malus*), the hypanthium is united to the carpels, the flowers are epigynous. The Rosoideae (e.g., *Rosa*) are perigynous, as are the Amygdaloideae (e.g., *Prunus*). The difference between these two sub-families lies in the fact that the Rosidae have several carpels and a fruit of the **cynorrhodon** type, while the Amygdaloideae have a carpel, most often single, and form a **drupe**.

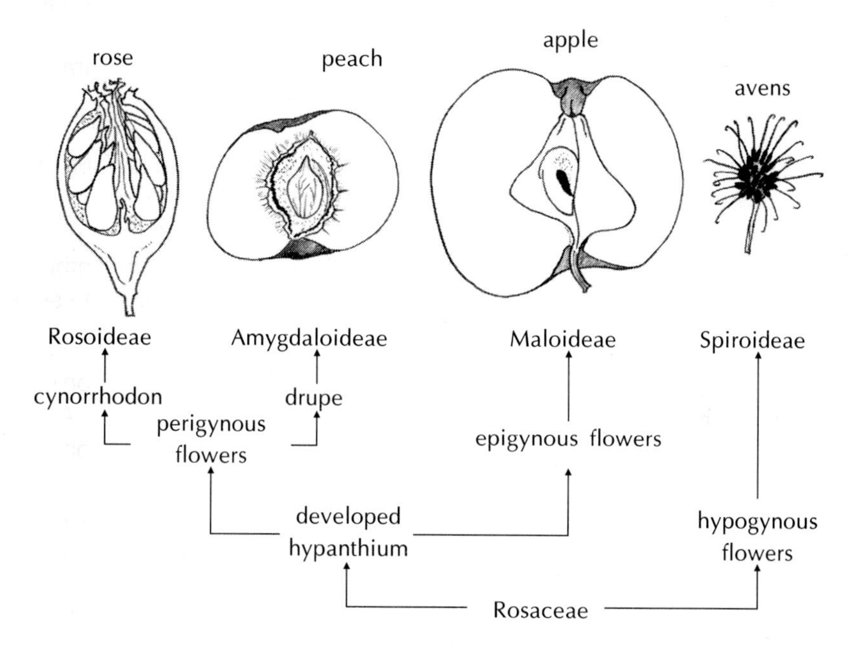

Fig. 3.6: Key to Rosaceae using hypanthium. An example of each sub-family is indicated below the corresponding diagrams. This figure does not represent the phylogeny of Rosaceae.

3.2.2. Carpels, Ovules and Phylogeny

Even though there is no general rule, two evolutionary "tendencies" can be suggested about the ovule of Angiosperms. The first is the transition from apocarpy to syncarpy and the second is the simplification of the crassicellular bitegmic ovule into a tenuinucellar unitegmic ovule.

a) Syncarpy, pseudomonocarpy

Apocarpy is not in fact very common; it is found in the Magnoliidae, Alismatales, and some Rosidae. The tendency to syncarpy has often been explained as an **adaptive advantage**, and the possibility of forming a **compitum** has been mentioned. A compitum allows the distribution of pollen tubes between different carpels and promotes **pollen selection**. This introduces a new level of **fertilization control**, and that is why the possession of a compitum is sometimes considered an evolutionary step as important as the transition from gymnospermy to angiospermy (Mulcahy, 1979).

Another advantage of syncarpy most often (and wrongly) neglected is the possibility of forming more complex fruits, i.e., those optimizing the dissemination of species.

> ***Remark:*** Undoubtedly for geometric reasons, syncarpous plants often have two to five carpels. It is possible that an excessive number of carpels would make union between carpels difficult.

It has also been supposed that the union of carpels produces first eusyncarpous gynoecia and then parasyncarpous gynoecia. Although this tendency seems to agree with some data, it is not always verified, since there are examples to the contrary: the Papaveraceae (Ranunculales), although among the "basal Tricolpates", are parasyncarpous. Moreover, the Orobanchaceae, effectively parasyncarpous, are not derived from the Scrophulariaceae (at least some of which are eusyncarpous), contrary to what would have been believed earlier.

Finally, in some genera there is a modification from the multicarpellary state to the **pseudomonocarpellary** state. The pseudomonocarpellary gynoecia appear monocarpellary and have only a single fertile carpel. The multicarpellary gynoecium begins its development normally but a single carpel results and the others abort. At best, these others are small carpels devoid of ovules. Among the Laurales,[3] the Lauraceae are pseudomonocarpellary, which is a derived character.

b) Ovules

In the Gymnosperms, the ovules have two integuments and it is likely, if not certain, that the ancestor of Angiosperms also had two ovular integuments. This means that, when unitegmic ovules are found in the Angiosperms, it is a derived character. The tenuinucellar unitegmic ovules exist particularly in the Orchidaceae, as well as in the Asteridae, in which this is a synapomorphy. The tenuinucellar unitegmic character causes an economy of space and thus allows ovules to occupy a single carpellary cavity (Endress, 1994), up to several millions in the Orchidaceae. In the case of plants that are mycotrophic

---

[3]Comprising the Lauraceae, Calycanthaceae, Chloranthaceae, and Monimiaceae.

(Orchidaceae, Rafflesiaceae) and bear seeds having no endosperm or perisperm that would cause a delay in development (the embryo directly uses the mycelial partner), the ovules are also tenuinucellar and sometimes even lack integuments (Gentianaceae).

c) Extreme reduction

The Santalales (generally parasites) and the family Balanophoraceae[4] have greatly reduced female reproductive parts: the gynoecium is very small, the **ovule aborts**, and an embryo sac forms in the thickened base of the gynoecium (Fig. 3.7). The plant thus produces no seed in the strict sense. This reduction appears in many Santalales, in a more or less marked manner, and is a derived character.

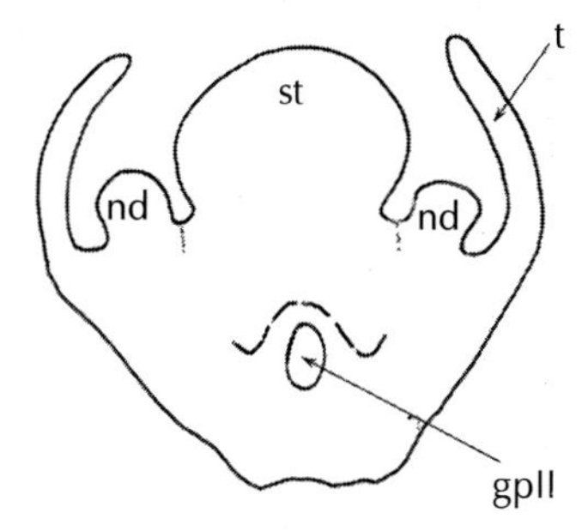

Fig. 3.7: Highly simplified diagram of a female flower of Viscaceae (Santalales) in radial section. The gynoecium comprises one or more secondary female gametophytes (embryo sac) that develop in the placenta (gpII), the ovules aborting early. The ovarian cavity does not form (its vestige is indicated by the broken lines). t, tepals; st, stigma, nd, nectariferous disc.

3.3. EVOLUTION AND OVERALL ARCHITECTURE OF THE FLOWER

3.3.1. From Spiral to Whorled Structure

The floral organs, as we have described them till now, particularly in *Arabidopsis thaliana*, are arranged in whorls, i.e., in concentric circles. However, this is not always the case, since they can also be arranged along a spiral (floral), forming a **spiral** floral structure (Fig. 3.8). A typical example is that of Ranunculaceae.

[4]Like the Rafflesiaceae, the Balanophoraceae are morphologically very different from other families of Santalales; moreover, the molecular data show atypical sequences. This is why these families are only hypothetically placed in the order Santalales on the basis of the parasitic mode of the life and ovarian reduction, a choice that we will follow here.

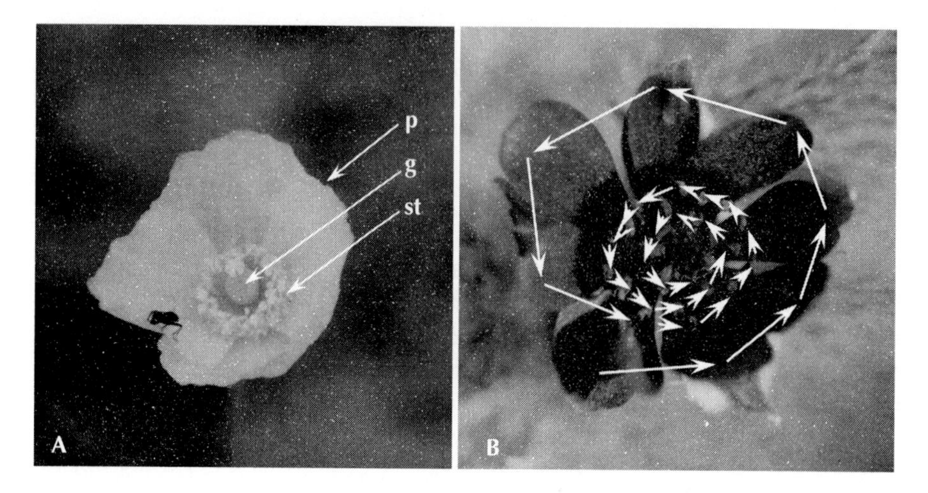

Fig. 3.8: Floral structures: (A) whorled and (B) spiral. The different whorls are indicated: p, petals, st, stamens, and g, gynoecium. The floral spiral is outlined by arrows showing the progress from one floral organ to the next.

From an evolutionary perspective, the spiral structure exists principally in the Magnoliidae, the paleoherbs, and the basal Tricolpates.

a) Theoretical perspective

As for the phyllotaxic spiral, the **angles of divergence** formed by floral organs conform to the results of Fibonacci (limit angle 137.5°) or those of Lucas (limit angle 99.5°). The existence of a recurrent limit angle in plants can be interpreted by formulating the hypothesis that an **inhibitor** produced by a primordium hinders the development of new primordia alongside (Richter and Schranner, 1978). Let us suppose, for example, that the concentration of the inhibitor, denoted as y, decreases exponentially in space and time. If it is placed in a circle, the initial primordium is at $\theta = 0$. The concentration of the inhibitor, y, at time t is written as follows:

$$y(t, \theta) = y_0 \cdot e^{-t\theta}$$

where t is the time, and y_0 is the concentration at $\theta = 0$ and at $t = 0$ (Fig. 3.9). In this case, y is minimal when θ is maximal (i.e., 180°), which means that the new primordium must establish itself at the position diametrically opposite to the first. Subsequently, the third primordium will form at the place where the total concentration of the inhibitor is the lowest. At first that place may be presumed to be at $\theta = \pi/2$, equidistant from primordia 1 and 2. This is not the case because by the time the third primordium forms the concentration of inhibitor emanating from the first primordium has diminished (an exponential decrease).

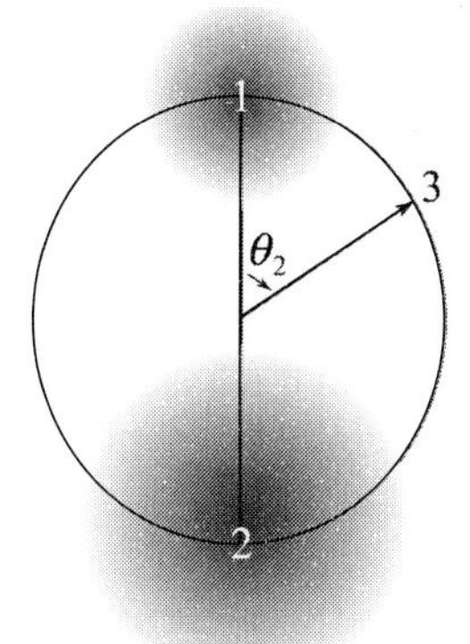

Fig. 3.9: Example of establishment of spiral structure. Primordium 3 develops where the concentration of the inhibitor (in grey) is minimal, i.e., not at 90° from primordium 1 or 2, but close to 70° (θ_2) from primordium 1 because the concentration of the inhibitor emanating from it has diminished by a factor 2 during the development of primordium 2.

Thus, it can be shown that if the concentration of the inhibitor from primordium 1 is reduced by a factor 2, then the third primordium forms at $\theta = \pi/2 - \ln 2/2$ (approximately 70°). Thus, gradually, the angle θ_n can be calculated at which the *n*th primordium is established, according to the formulas of recurrence:

$$\theta_n = \arg_\theta \min \Gamma_{n-1}(\theta), \text{ where } \Gamma_{n-1}(\theta) = \sum_{p=1}^{n=1} y_p(n, \theta)$$

$$y_n(t, \theta) = y_0 e^{-(t-n+1)\,\varphi_n(\theta)}, \text{ where } \varphi_n(\theta) = \theta - \theta_n \text{ or } 2\pi - \theta + \theta_n$$

where y_n is the concentration of the inhibitor emanating from primordium *n*. For example, for $n = 4$, we find $\theta_4 = 267.6°$. The angle of divergence, a quantity that is more useful than θ_n itself, is the difference $\theta - \theta_{n-1}$. We find, always in the context of our example, successively 180°, 109°, 162°, etc. The angles approach a limit value, a result of Fibonacci.

b) Controls and development

The angle of divergence frequently observed is often easy to calculate. Although the model is attractive, it is not a proof for the existence of the inhibitor, which still remains to be identified. Supposing that this mode of control "by inhibition" of establishment of floral organs does exist in flowers with a spiral structure, that shows that the cadastral control involving the flower has probably changed from the central Tricolpates (Vitales and Caryophyllales, at the base of the central Tricolpates, "already" have a whorled structure), since there are whorled flowers with an alternate or opposite arrangement of organs.

Besides, the control goes further, i.e., up to the **temporal level** (timing of establishment of primordia). In the case of whorls, all the organs of a single whorl are initiated simultaneously. In the case of "ideal" spiral structures, the floral organs are established successively, one after the other. Still, it must be remembered that there are organisms (e.g., *Austrobaileya* sp.) in which there are not one but several (five) floral spirals (Fig. 3.10). In this case, the five primordia, on their respective spirals, develop at the same time, which could lead to confusion, making it look like a whorled flower. Conversely, a whorled flower could show temporal gaps in the initiation of primordia, especially when there are several whorls of organs of a similar nature, like stamens. In this case, the organs concerned developing first are the innermost (centrifugal development) or the outermost (centripetal development). It could happen, for steric reasons, that the organs formed last are smaller than those formed earlier. Finally, the flowers with a whorled structure sometimes bear stamens in **tufts** (Hypericaceae), each tuft being arranged as a single stamen would be. These tufts come from the **subdivision** of stamen primordia, so that the stamens, once they are formed, remain united at the base (Fig. 3.10).

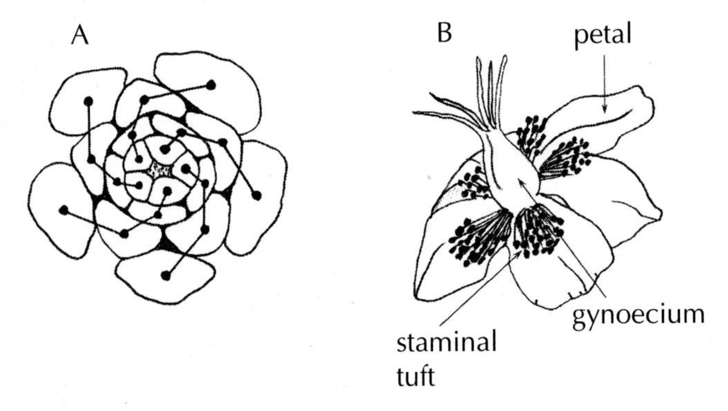

Fig. 3.10: Examples of floral organization. (A) Flower with helical structure with five floral spirals of Austrobaileyaceae (only the positions of primordia are shown in the illustration). (B) Flower of a Hypericaceae partly flattened, showing the staminal tufts.

c) Atypical structures

The term "atypical" here refers to flowers that seem "chaotic" and are thus described as **irregular**, i.e., neither whorled nor helical. This is found in Magnoliidae and sporadically in other groups.[5]

3.3.2. From Actinomorphy to Zygomorphy

A flower is said to be **actinomorphic** when its structure has axial symmetry. It is also said to be **polysymmetrical**, because such a flower has several planes

[5]Some authors have related this character with short plastrochrones, or the absence of a perianth (Endress, 1994), but this relation is far from being automatic.

of symmetry, the intersection of which is a straight line: the **axis** of symmetry. **Zygomorphic** or **monosymmetrical** flowers, on the other hand, have only one plane of symmetry. In this case, the flower has two sides, on either side of a single plane. This is why we speak of **bilateral** symmetry (Fig. 3.11). The bilaterality often implies **dorso-ventralization** of the flower, i.e., a floral differentiation along the plane of symmetry. The flower could thus have two **lips** (e.g., Lamiaceae), one dorsal, the other ventral. There are a very large number of zygomorphic structural planes possible, which correspond to different "ways of obtaining" the bilateral symmetry from an actinomorphic flower.

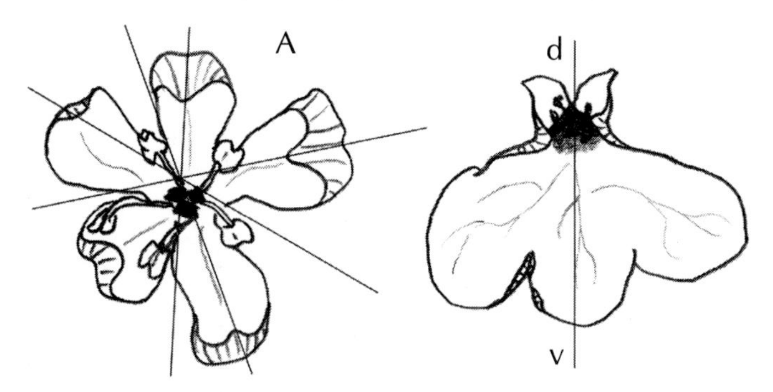

Fig. 3.11: Floral symmetry: examples of an actinomorphic flower of Anacardiaceae (A) and a zygomorphic flower of Acanthaceae (B). The axis of symmetry is indicated in (B), while (A) shows the paths of four possible axes of symmetry. The bilateral symmetry leads to the differentiation of a dorsal side (d) and a ventral side (v).

Remark: The classification of flowers into actinomorphic or zygomorphic is nevertheless a simplification. First of all, actinomorphy strictly speaking (more than two planes of symmetry) is distinct from **bisymmetry** (two planes of symmetry). With respect to zygomorphy, there are at least two sub-types: median zygomorphy (bilateral symmetry) and **diagonal zygomorphy**, in which the plane of symmetry is not entirely vertical. Finally, some flowers cannot be classified according to these categories, in the sense that they are asymmetrical, because they have a spiral structure (see above), or even because the floral organs, apart from the perianth, have only left or right morphs. The last case is referred to as **enantiomorphy**.

Zygomorphy must be understood in the context of modes of pollination. Whereas actinomorphic flowers are generalists, i.e., have a range of possible pollinators, zygomorphic flowers have a restricted number of pollinators, in this case pollinators that are small enough to slip into the corolla, which is frequently tubular. In other words, zygomorphy involves some **floral specialization**. Till now, it has been supposed that zygomorphic flowers are **less**

variable, especially in size, because any modification in a specialized flower would alter the close correspondence between flower and pollinator and lead to a drop in the pollination efficiency, and thus in reproductive success. This was corroborated by observations on flower size (Wolfe et al., 1999): measurements taken on 31 species of Angiosperms showed that bilateral flowers are significantly less variable in size than actinomorphic flowers (Fig. 3.12).

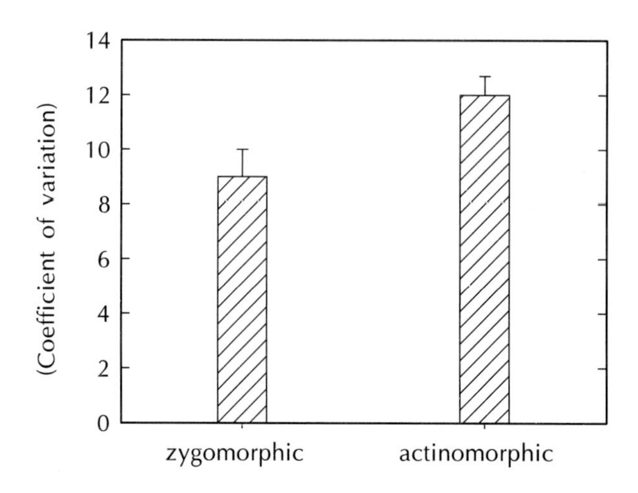

Fig. 3.12: Coefficient of variation of flower size (coefficient of variation = standard deviation/ mean) observed in 31 species of Angiosperms with zygomorphic or actinomoprhic flowers (modified from Wolfe et al., 1999).

There exist predominantly zygomorphic taxa in which some members are actinomorphic, which suggests a **reversion** towards actinomorphy (by loss of genetic function, like *cycloidea*, see below). The return to actinomorphy could be an adaptive advantage in some difficult environments with few pollinators, such as an alpine environment (e.g., *Ramonda* sp., Gesneriaceae). Nevertheless, this aspect cannot be taken out of the floral context: the return to actinomorphy is accompanied by modifications that allow easy access to pollinators (the *cycloidea* mutants of snapdragon are sterile). This return therefore implies not simply a loss of function but also a change in the depth of the tubular corolla, as is probably the case in *Ramonda*.

a) Genes and symmetry

Actinomorphic mutants of plants with ordinarily zygomorphic flowers have been known for a long time: for example, the first description of actinomorphic specimens of common toadflax (*Linaria vulgaris*, Scrophulariaceae) was given in the 18th century by Linnaeus, who named these mutants *Peloria*. Mutants of this kind are therefore called **peloric**. In the 19th century, mutants of snapdragon

(*Antirrhinum majus*, Scrophulariaceae) intrigued Darwin, and particularly a famous peloric variety of type 3, with 6 stamens, called *cycloidea* (Fig. 3.13). The wild type had 4 stamens and a minute staminode. In place of that staminode, the mutant *cycloidea* had a stamen, which Darwin considered an ancestral character (**atavism**). Genetic studies subsequently showed that the locus *cyc* considered conventionally bore in fact two loci, corresponding to the genes *cyc* and *rad* (*radialis*). The protein Cyc, which has no homology with known proteins, is expressed on the dorsal (adaxial) side of the flower, creating the zygomorphy, undoubtedly slowing the development of adaxial primordia to some extent. However, Cyc does not act alone, and the acquisition of zygomorphy also requires the genes *rad*, *dich* (*dichotoma*), and *div* (*divaricata*, which is expressed abaxially). It is now known that the genes *rad* and *cyc* are expressed adaxially, the gene *div* abaxially. The expression of *div* is inhibited adaxially by *cyc* and *dich* (Hudson, 1999).

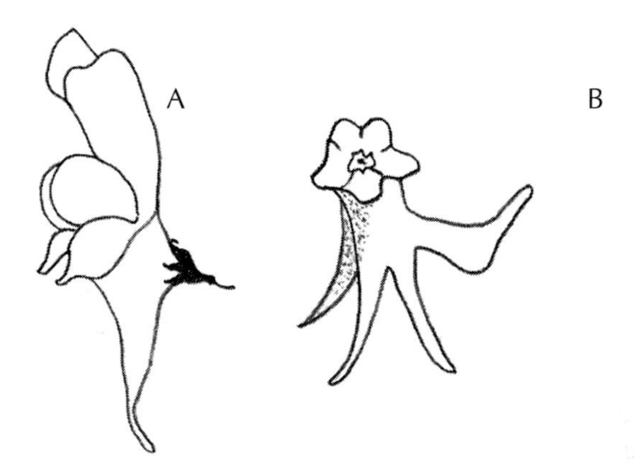

Fig. 3.13: Appearance of zygomorphic wild types (A) and actinomorphic mutants (B) of snapdragon flowers (*Antirrhinum majus*) in external view.

The first step in this cascade of genetic regulation is the adaxial expression of *cyc*, which suggests that it responds to the orientation of the inflorescence axis and bract. The mutants *centroradialis* of snapdragon produce flowers that do not have dorso-ventralization, from the inflorescence meristem. Moreover, ablation of the inflorescence meristem zone adjacent to the floral meristem led to the formation of an actinomorphic flower. On the other hand, if this ablation is linked to an application of auxin in the same place, a zygomorphic flower forms (Hudson, 1999). It therefore seems probable that it is the auxin from the inflorescence meristem that triggers the expression of *cyc* and thus leads to the establishment of floral zygomorphy (Fig. 3.14).

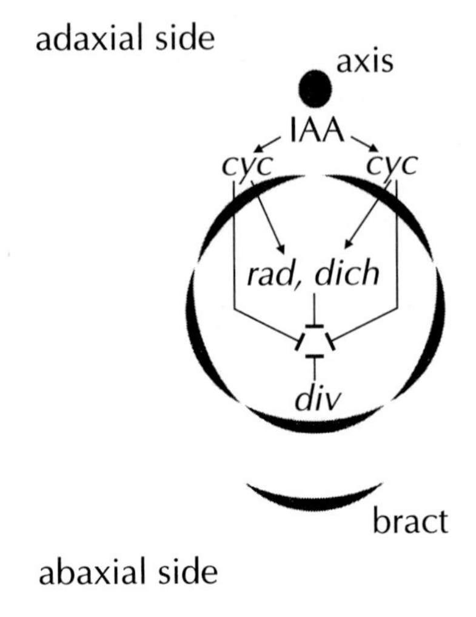

Fig. 3.14: Molecular determination of floral symmetry. Arrows, activation; bars, inhibition; IAA, auxin (indole-3-acetic acid). The genes indicated are described in the text.

b) Phylogenetic perspective

Zygomorphy is found in **many families**: Fabaceae, Lamiaceae, Scrophulariaceae, Plantaginaceae, Bignoniaceae, etc., i.e., mostly among the Asteridae.[6] One of the first observations about the phylogenetic aspects of symmetry is that, since there are numerous peloric mutants in nature, there are no inverse mutants (zygomorphic mutants of flowers that are actinomorphic). This undoubtedly suggests that though the zygomorphic-actinomorphic transition can be envisaged, the actinomorphic-zygomorphic transition is much more difficult. This is not necessarily compatible with several independent appearances of zygomorphy.

Phylogenetic analyses, supposing the equality of probabilities of two transitions, have primarily been realized in the Asteridae, by means of phylogenetic trees established as detailed earlier. If the hypothesis of ancestral actinomorphy is formulated, that implies three independent transitions towards zygomorphy. The inverse hypothesis (the zygomorphy is ancestral) leads to four transitions towards actinomorphy. Studies done on more complete trees also show that the first hypothesis is the most economical in terms of transitions

[6]There are zygomorphic flowers in some families that do not belong to Asteridae (e.g., Brassicaceae—see above, Fabaceae, Ranunculaceae, etc.). That said, the genetic origin does not seem to be the same: the bilateral symmetry of Brassicaceae flowers of type 4 results from the expression of a gene direced towards the control of a number of organs (*perianthia*), while in Asteridae there is a gene fundamentally linked to asymmetry, without direct effect on the number of organs (*cycloidea).* This hypothesis remains to be confirmed.

(Donoghue et al., 1998). If we now suppose different probabilities of transition, these probabilities are still unknown and need to be fixed. In fact, we have already seen that the transition towards actinomorphy seems inherently easier from a genetic point of view; however, actinomorphy does not necessarily persist in an initially zygomorphic population, because the "generalist" pollinators visiting the actinomorphic flowers are less effective. In any case, the hypothesis of ancestral actinomorphy of Asteridae, linked to several independent transitions towards zygomorphy, is presently agreed upon (Fig. 3.15).

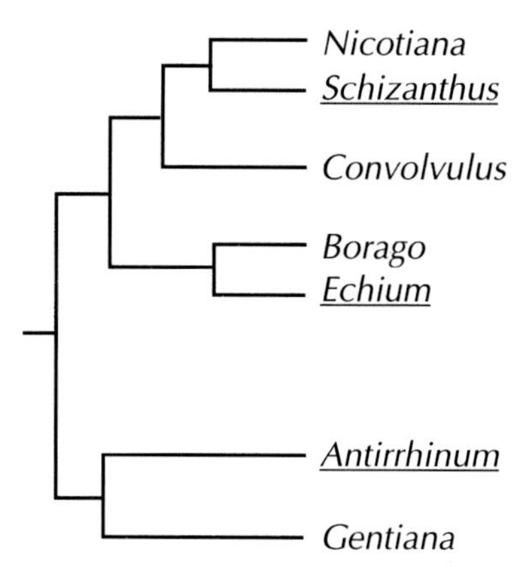

Fig. 3.15: Cladogram of Asteridae considering actinomorphy as ancestral and leading to at least three independent appearances of zygomorphy (genera underlined) (modified from Donoghue et al., 1998).

Reversions from zygomorphy towards actinomorphy exist in Lamiales in particular, where they seem to be associated with the acquisition of a structure of type 4, as in *Plantago*. It is presently supposed that this results from the fusion of two adaxial petals, as is suggested by the existence of a relatively widened adaxial petal in *Veronica*, a similar genus (Donoghue et al., 1998). Moreover, this does not mean that the expression of a gene of the *cyc* type will not be normal in *Plantago*. The actinomorphy does not necessarily result from a loss of function but could also arise from the superimposition of a process of union.

c) Design of floral organization and bilateral symmetry

The Asteridae essentially have flowers of type 5. There are thus four possibilities of zygomorphic organization (Fig. 3.16). Zygomorphy implies, as we have said, a floral dorso-ventralization. Taking into account the floral orientation between the bract and the inflorescence axis, the corolla can simply be

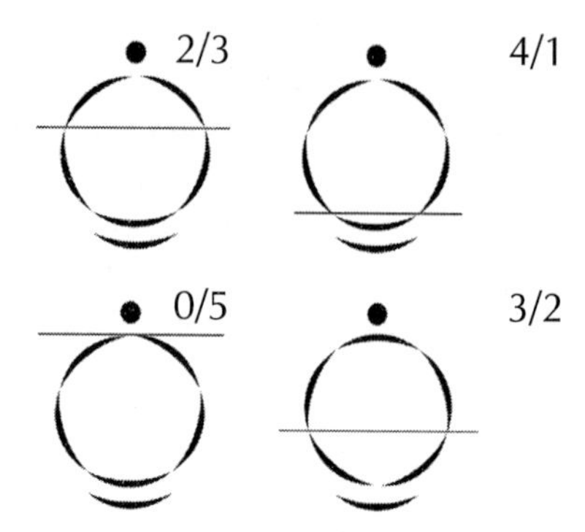

Fig. 3.16: Various modes of dorso-ventral sectioning of the corolla of zygomorphic flowers. The adaxial side is turned towards the axis (point) and the abaxial side towards the bract (crescent).

sectioned into dorsal (i.e., adaxial) and ventral (i.e., abaxial) parts in three ways: 4/1, 2/3, or 0/5. These three patterns exist in nature, respectively in honeysuckle (*Lonicera caprifolium*, Caprifoliaceae), dead-nettle (*Lamium* sp., Lamiaceae), and germander (*Teucrium* sp.). The 5/0 section does not seem possible, because it would be based on suppressing the abaxial petal, or dividing it into two parts. Let us remark, finally, that although we analyse zygomorphy on the basis of the petal whorl, that does not mean that the sepals are not affected by dorso-ventralization, even though the latter is certainly less spectacular than in the case of the corolla.

Still, 3/2 architecture is seen, for example in rhododendron (Ericaceae). This results from the existence, in this case, of the inversion of orientation of the entire floral primordium, i.e., at their formation, the bract faces a sepal and not a petal, as in the majority of flowers. This configuration is found in the Rosidae, where the zygomorphic flowers of Fabaceae also have a 3/2 cutting, the **carina** (keel) (see Chapter 4) resulting from the union of two abaxial petals. We must see these particular cases as "special" architectures, derived among the taxa belonging to Ericaceae and the Fabales-Rosales-Cucurbitales-Fagales group, where we find essentially actinomorphic flowers. The Polygalaceae, a family close to the Fabaceae and belonging to the Fabales, also have zygomorphic flowers, but the orientation is respected (Fig. 3.17). Besides, dorso-ventralization in this case affects mostly the calyx, cut into 1/(2 + 2).

Among the Monocotyledons, the Orchidaceae have zygomorphic flowers of type 3, in which the corolla architecture is 2/1. The abaxial petal is differentiated into a **labellum**. Here also, there is a problem in floral orientation, since the bract does not face a sepal but a petal (see Fig. 4.5). In this family, unlike in rhododendron, the floral primordium undergoes a **rotation** of 180° in the

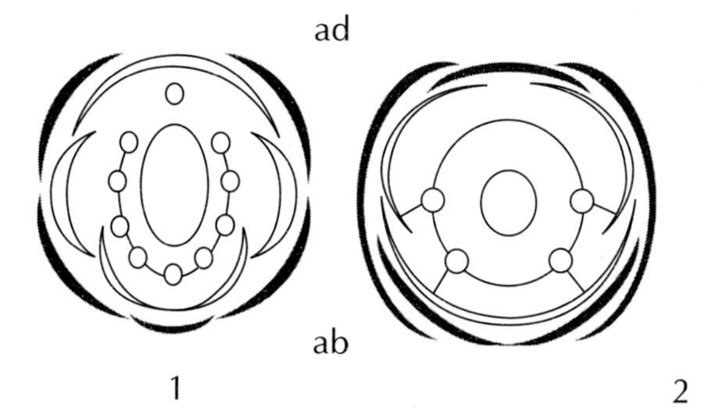

Fig. 3.17: Comparison of the floral organization of Fabaceae and Polygalaceae using floral diagrams of a Faboidea (1) and a Polygalacea with four stamens (2). The adaxial side (ad) is as always oriented towards the top. The stamens are connate by the filament and adnate to the petals in the Polygalaceae, while they are only connate in the Fabaceae. One petal is highly developed in the two families: the adaxial standard in the Fabaceae and the abaxial petal in the Polygalaceae. Nevertheless, they have the same embryo structure, the floral primordium of the Fabaceae being resupinated, bringing the adaxial side close to a petal and not to a sepal.

course of its development. On the other hand, in the Commelinaceae, there is no reversal and the zygomorphic flowers, where they exist, have a 1/2 architecture.

3.3.3. "New" Floral Organs

a) From tepals to "petals"

In the Nympheales, Magnoliales, and Ranunculaceae, there are **tepals** present. Among the Ranunculales, Papaveraceae and Berberidaceae show a **differentiation** of tepals into petals and sepals. The Proteales and Vitales, "basal" Tricolpates, also have whorls differentiated into calyx and corolla. In the monophyletic group Caryophyllaneae (*Caryophyllanae*), the Caryophyllales have flowers fundamentally with **tepals**, but with an architecture that varies widely from one family to another. In this sense, the evolution of the floral structure of Caryophyllales is remarkable (Fig. 3.18).

The Caryophyllaceae (e.g., silene, carnation) generally have floral parts apparently differentiated into petals and sepals. But this only appears to be so: the outermost whorl in fact has 4 or 5 tepals and is followed by 4 or 5 petaloid stamens, ordinarily called "petals". These are followed by 4 to 10 stamens and then, centrally, the carpels. The Phytolaccaceae (e.g., *Phytolacca*) have no staminal differentiation and the flowers have 5 tepals followed by 10 (or more) stamens. In the Nyctaginaceae (e.g., four o' clock, bougainvillea), the sepaloid (or petaloid) bracts support a flower with 5 tepals (the perigonium is also

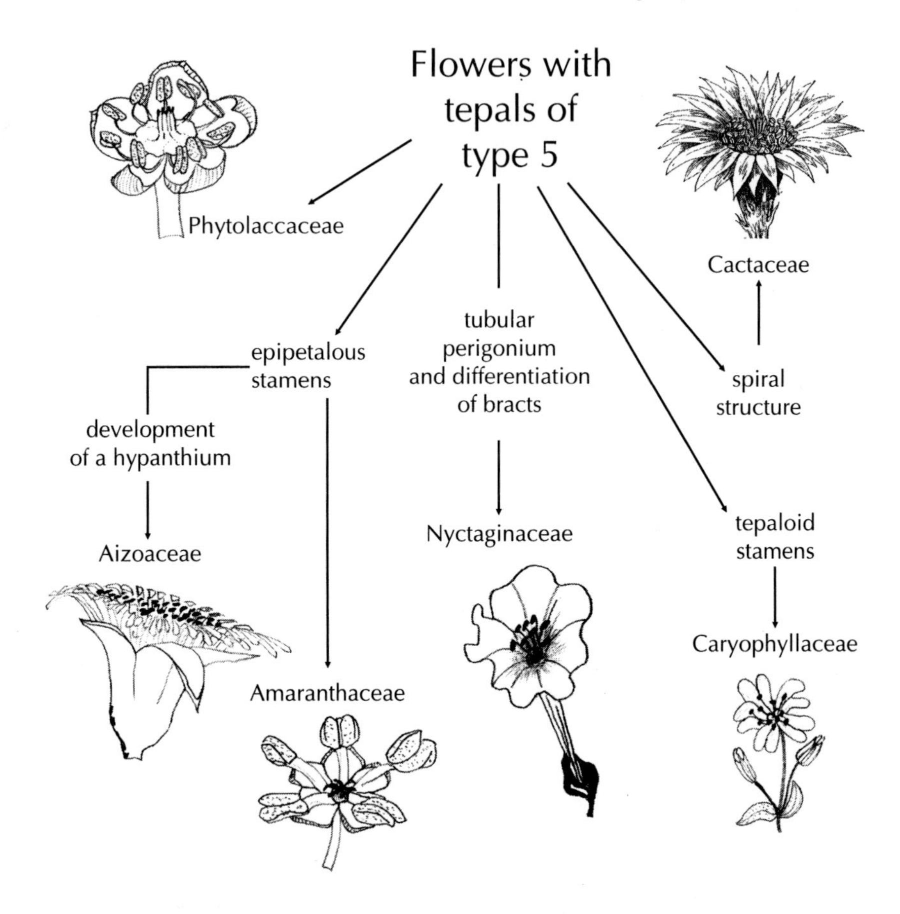

Fig. 3.18: Key to Caryophyllales using the characters of the perigonium. This figure does not represent the phylogeny of the group.

differentiated into a **tubular** zone and a **flattened** zone), 5 stamens, and 1 carpel. Flowers of the Amaranthaceae (including the earlier family "Chenopodiaceae") have a structure comprising 3 to 5 tepals, 5 **epitepalous** stamens, and 2 to 3 united carpels. The Portulacaceae[7] have a similar floral architecture (epitepalous stamens). A **hypanthium** is noted in the Aizoaceae (including the Mesembryanthemaceae). Besides, these flowers have 5 tepals (connate or not) and sometimes numerous stamens, the outermost of which are thus petaloid. The Cactaceae have also a hypanthium and, what is remarkable, a spiral arrangement of floral parts. The tepals change gradually in

[7]Including the Basellaceae and Didieraceae. However, the monophyly of this group remain to be specified. For the present, and before more precise dendrograms are drawn, the Portulacaceae are left as such, on the basis of the association of the flower with a pair of sepaloid bracts.

appearance as one goes into the flower: tough towards the exterior and looking like petals towards the interior. The stamens are numerous.

Thus, we see that from an ancestral structure with **tepals**, the Caryophyllales, clearly monophyletic (we will not here survey the synapomorphies), are organized into families characterized by the acquisition of highly varied derived floral architectures, which sometimes results in a differentiation of parts of a similar nature, like stamens, into petaloid parts and stamens. Also, the importance of bracts should be noted in the Nyctaginaceae; they serve the same function, notably protection, that sepals serve in other taxa. Curiously, some plants with true sepals and petals, such as *Thunbergia* (Acanthaceae), show a considerable reduction of the calyx in favour of the bracts, in a way similar to the Nyctaginaceae (Fig. 3.19). This is a remarkable example of **convergence**.

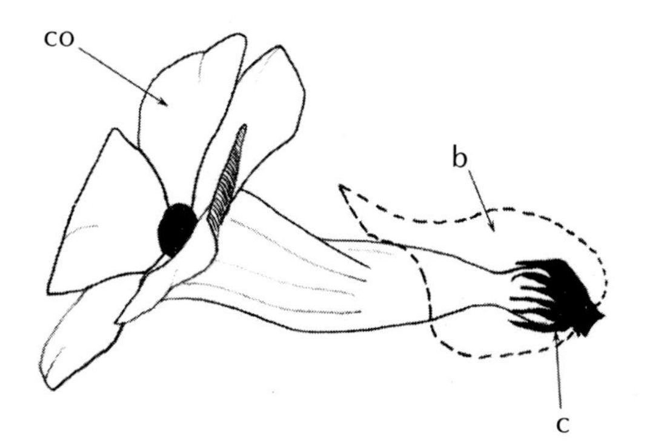

Fig. 3.19: Appearance of a *Thunbergia* flower (Acanthaceae) showing the significant reduction of sepals (c) in favour of the bract, the location of which is indicated by the broken line (b); co, corolla.

b) Floral appendages: corona

The Asclepiadaceae have a remarkable and highly complex floral architecture, particularly because of the presence of specific structures, which makes this group an unavoidable subject of study in botany. They are in fact a co-family of Apocynaceae (e.g., periwinkle), which is to say that their floral architecture is derived.[8]

Floral architecture (*Asclepias*)

The floral formula, unvarying among the Asclepiadaceae (around 2900 species) and of the type *, 5, 5, 5, 2, is typical of Apocynaceae. On the other hand, this co-family has additional parts, between petals and stamens, forming the

---

[8]This means that, in terms of nomenclature, the word Asclepioidae should be used, not Asclepiadaceae.

corona (Fig. 3.20). Moreover, stamens and carpels are associated and form the **gynostegium**. Two adjacent stamens deform on contact to form the **rail**. Just above the rail, a cavity of complex shape is filled by a secretion from the style (it is thus a homologue of a stigmatal secretion) slightly before anthesis and this secretion, once solidified, forms the **clip**. The clip is attached to the **pollinia**[9] of two stamens that enclose it with two arms. Each stamen thus produces two pollinia joined to two different clips. The entire clips/pollinia system constitutes the **pollinarium**. The gynostegium results from the terminal suture of the styles of two carpels, the whole being more or less adnate to the stamens, thus forming a **stylar head**. The common stigma is curiously pentapartite because the stamens have served as a "mould" during development. When the secretion of the clip detaches, the corona develops. It is composed of five parts, alternate with the petals, themselves organized into **cup** and **horn**. Such an architecture confers some specialization with respect to pollinators and remarkable pollination efficiency. From one species to another, there are variations on this plan (typical of the genus *Asclepias*) linked to modifications of ornamentations (trichomes, coronary protrusions, etc.). Finally, the Asclepiadaceae frequently have a strong fragrance (e.g., *Stephanotis*).

Origin of the corona

It has still not been established with certainty whether the corona is an appendage of the corolla or of the stamen. In particular, *in situ* hybridization experiments might reveal where the genes of functions A, B, and C are expressed (see Chapter 2). In some species lacking a corona, however, there are diverticuli

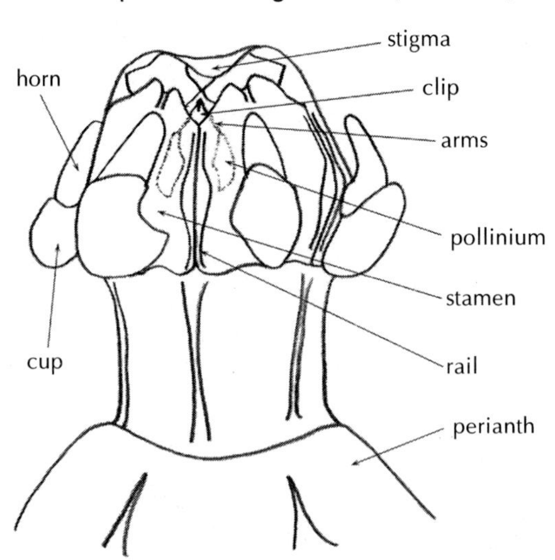

Fig. 3.20: Simplified diagram of a part of the flower of *Asclepias* showing especially the organization of fertile parts.

[9]The Asclepiadaceae effectively bear pollinia, like the Orchidaceae.

of petals, which could be prefigurations of the corona, suggesting that it could originate from the corolla.

Phylogenetic perspective
In the Apocynaceae, in general, the stamens are adnate to the petals. There are sometimes nectaries, alternating with the two ovaries. The derivation of the flower of Asclepiadaceae from an ancestor of the entire group Apocynaceae thus supposes modifications, such as the doubling of the whorl of petals, formation of the corona associated with the stamens, and the taking over of nectar production when the nectaries disappear (this is not always the case). Besides, the stamens are united with the style and, especially in *Asclepias*, lose one pollen sac per pollen locule (the remaining sac produces a pollinium). The flower finally loses its tubular organization, so that the petals, instead of forming a tube, are directed towards the base of the flower. Genetic studies remain to be done to specify the modalities of establishment of floral parts, and the homologies between parts. In any case the corona is a virtual floral **innovation**, even though a corona also exists in the Passifloraceae (and, among the Monocotyledons, in the Amaryllidaceae (narcissus)). This is a convergence. Moreover, the corona of Passifloraceae has a filamentous appearance, very different from that of the Asclepiadaceae.

3.4. EVOLUTION OF THE FLOWER AT THE SCALE OF THE WHORL

It is not possible, in a subject of this scope, to be exhaustive. The evolution of the flower at the scale of a whorl, starting from the synapomorphies of taxa, could effectively be developed for each of the 52 botanical orders! This is why we limit ourselves to drawing out the important ideas, illustrating with one or two orders that are typical or "inevitable" in terms of their peculiarities or their occurrence. There is no section specifically devoted to the **union** of floral parts, because they are implied in highly diverse architectures. We will simply indicate that the union of **petals** is found particularly in the Asteridae, still called Sympetals for this reason (*Sympetalae*). We will cite, as a matter of interest, the Ericales (Ericaceae, Primulaceae, etc.), Solanales (Solanaceae, Convolvulaceae, etc.), Gentianales, Lamiales, and Asterales.

3.4.1. Modifications

a) General points

The term "modification" has been deliberately chosen because it is vague and potentially contains several possibilities: e.g., disappearance or reduction, hypertrophy (overdevelopment), transformation of certain parts of a whorl. The case of **disappearance** particularly concerns zygomorphic flowers in which bilateral symmetry imposes an underdevelopment of adaxial parts (see above), which could go as far as their omission (e.g., stamen of Lamiaceae). Disappearances could be accompanied by the transformation of petals or sepals into bract-like parts (e.g., Poales, Chapter 2). The parts of the perianth may not exist at all, in a derived mode (e.g., Piperaceae, Chloranthaceae). An important example of **transformation** is that of staminodes and pistillodes, i.e., the transformation respectively of fertile stamens and carpels into sterile organs implicated in the production of nectar or more generally in the attraction of pollinators. Many groups have staminodes, such as the Rosaceae or Zingiberaceae (see section 3.4.3), as well as zygomorphic Dicotyledons. Indeed, as we have said, bilateral symmetry imposes a modification of the development of adaxial parts, and particularly of the adaxial stamen of flowers of type 5, such as the Lamiales, which we will use as an example.

b) From fertile parts to sterile parts

Staminodes (Lamiales)

In the Lamiales group, it is frequently remarked that, because of zygomorphy, the androecium is didynamous (organized into (2 + 2) stamens + 1 staminode) or reduced to 2 stamens + (2 + 1) or 2 staminodes. For example, the Scrophulariaceae and Plantaginaceae have an adaxial staminode in only 30% of cases (Endress, 1994), while the Bignoniaceae often have one. But it is not possible to give general information because, even within a single family, when there remain only 2 stamens, they are either the superior or the inferior (except in the Lentibulariaceae, in which they are always the inferior).

In the Plantaginaceae (which presently include the Callitrichaceae as well as *Veronica, Linaria, Antirrhinum*, and *Digitalis*, which earlier belonged to the Scrophulariaceae—see box), there are most often 4 stamens, the adaxial stamen being present as a staminode in *Penstemon* and some other genera. The corolla is composed of 5 petals or sometimes 4, following the fusion of two adaxial petals (see above). The androecium is similar in the Scrophulariaceae[10] (e.g., *Verbascum, Scrophularia*) or the Orobanchaceae[11] (*Pedicularis, Euphrasia, Orobanche, Striga, Melampyrum, Rhinanthus*).

The Bignoniaceae have 5 united petals, 4 didynamous stamens, or just 2, and 1 staminode that may be protruding. In the case of *Jacaranda*, the staminode, which clearly emerges from the corolla, has glandular trichomes.

Moreover, although the lateral disposition of stamens was not clearly marked (limited to a curvature of the filament) in the preceding families, it is

---

[10]One character of Scrophulariaceae is the confluence of pollen sacs towards a common opening.

[11]Synapomorphies: parasitic mode of life, production of orobanchines (phenolic glycosides).

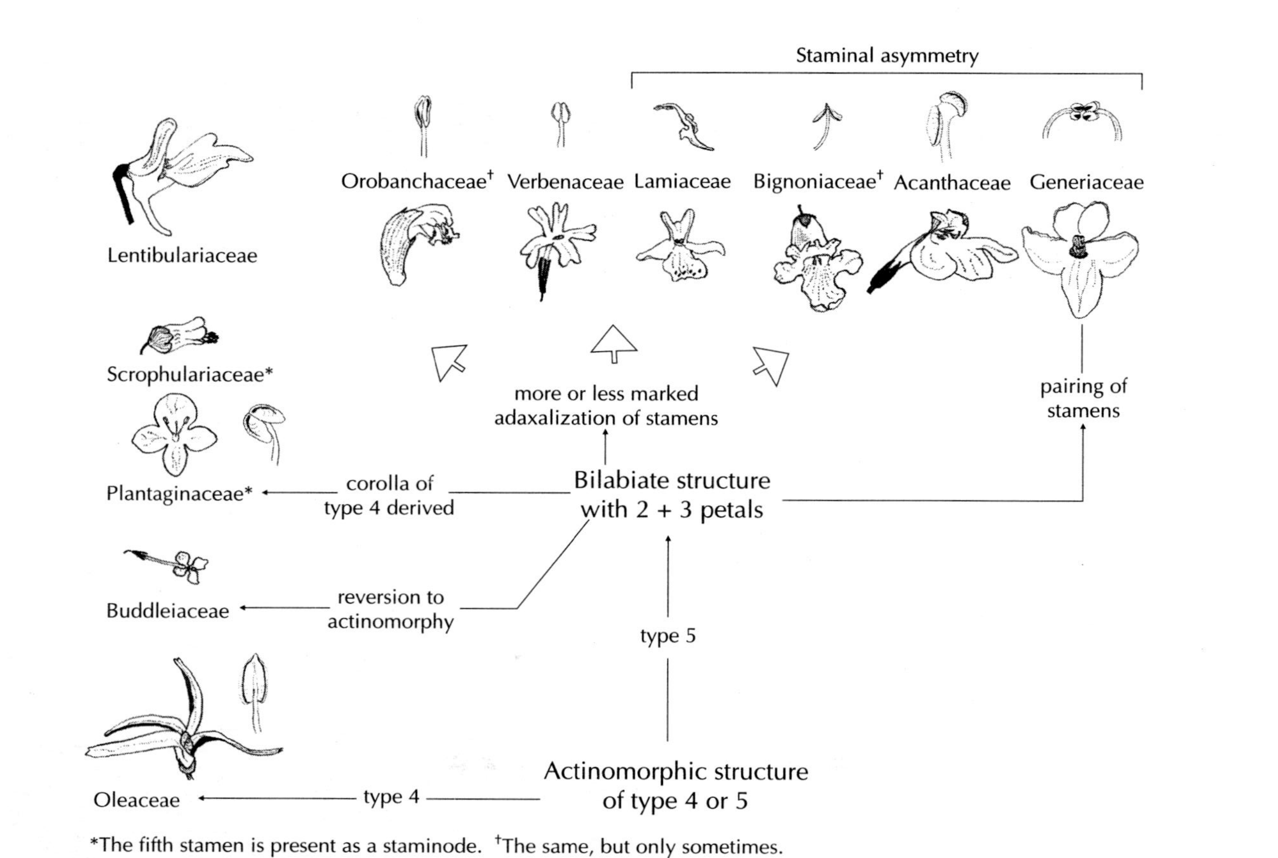

Fig. 3.21: Diagram showing the diversity of Lamiales, using notably staminal characters and symmetry. This representation is not a phylogeny.

The Scrophulariaceae, a Paraphyletic Group

The Scrophulariaceae, in the classical sense of the term, comprised a set of species whose common characters are difficult to see, partly because some species are morphologically close to the Lamiales but without the typical traits of families of this order were included in the group. The Scrophulariaceae contained genera as different as *Mimulus, Pedicularis,* and *Verbascum*. The absence of characteristic traits of this family (synapomorphies) gave rise to the hypothesis of the paraphyly of Scrophulariaceae. Studies undertaken from 1995 onward on the systematics of Scrophulariaceae effectively corroborated the hypothesis of "non-monophyly" (Olmstead et al., 1995). Subsequent studies using nucleotide sequences of plastidial genes (*rbcL, ndhF*, and *rps2*) led to a break up in the Scrophulariaceae (Olmstead et al., 2001). The Scrophulariaceae *s.s.* are composed of genera such as *Verbascum, Buddleia*, and *Scrophularia* (Fig. 3.22). *Antirrhinum, Digitalis, Globularia, Plantago*, and *Veronica* are placed among the Plantaginaceae. In morphological terms, we can put veronica with plantago because the corolla with four lobes results from the union of two adaxial petals (see above). This observation is corroborated by the phylogeny, which associates *Plantago* and *Veronica*. With respect to the tree as a whole, three genera are remarkable:

— *Buddleia*: If the Buddleiaceae were artificially raised to the rank of family, that would result in the elevation of other groups to this rank, with less reliable confidence (*bootstrap* percentage). The genus *Buddleia* thus seems included in the Scrophulariaceae *s.s.* along with *Scrophularia* and *Verbascum*, even though its floral architecture of 4 corolla lobes distances it morphologically from the latter two genera.

— *Mimulus*: The place of this genus is uncertain, but it is highly probable that *Mimulus* does not belong to the Plantaginaceae. Perhaps it will be necessary to consider the creation of the family Mimulaceae.

— *Calceolaria*: This genus of plants with well-known relationship actually falls outside the Scrophulariaceae *s.s.* or even the Plantaginaceae, and if *Calceolaria* is artificially associated with either of those groups the trees obtained are less reliable (*bootstrap* test). Thus, it seems likely that a family Calceolariaceae will have to be recognized, including the genera *Calceolaria, Jovellana*, and *Porodittia*.

Note that the Orobanchaceae henceforth comprise the genera *Rhinanthus, Pedicularis*, and *Striga*, the parasitic character being a synapomorphy of this family among the Lamiales. Finally, the aquatic plants *Hippuris* and *Callitriche* seem clearly included among the Plantaginaceae, resulting in the elimination of Hippuridaceae and Callitrichaceae.

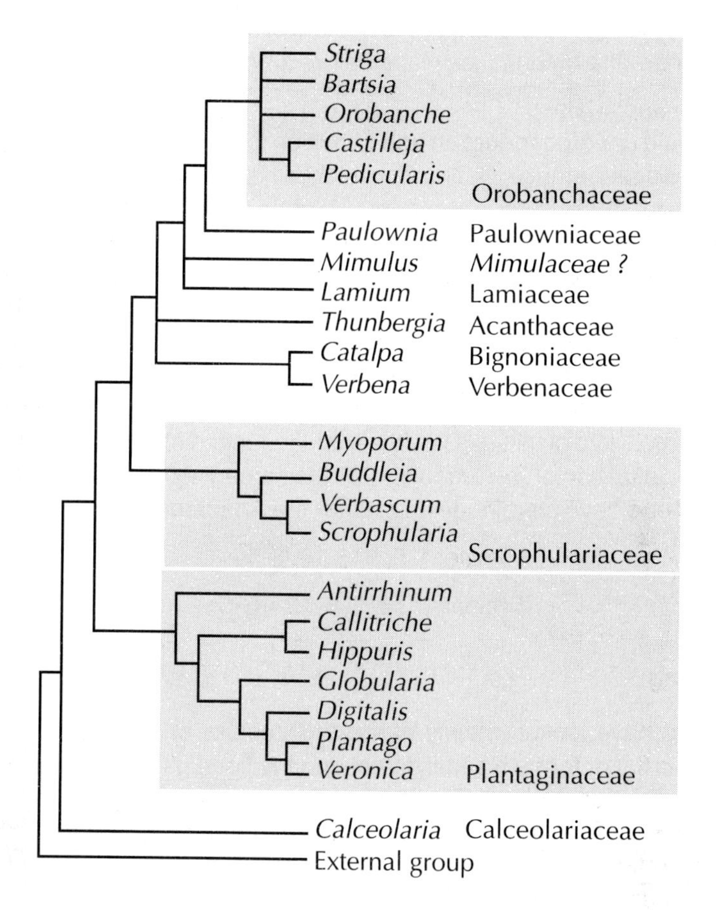

Fig. 3.22: Simplified consensus phylogenetic tree of major groups of Lamiales, constructed from plastid gene sequences *rbcL, ndhF*, and *rps2*. Note the breaking up of "conventional" Scrophulariaceae with especially the individualization of Plantaginaceae. The genus *Mimulus* is not very close to the other genera of Lamiales, which could lead to the creation of the family Mimulaceae (the question mark underlines that this hypothesis remains to be confirmed). *Calceolaria* branches out at the base of the tree and the constitution of the family Calceolariaceae is envisaged (modified from Olmstead et al., 2001).

clear in the Bignoniaceae, one pollen locule being smaller than the other (Fig. 3.21).[12] The lateral disposition is still more striking in the Acanthaceae, with profoundly asymmetrical stamens. This asymmetry is linked to the presence of **staminal horns** (and trichomes) The horns are appendages of the anthers, directed toward the base, so that the passage of the pollinator leads to the opening of the dehiscence pore. The curvature is also more pronounced, to

[12]However, there is a functional asymmetry of pollen locules in Lamiaceae such as sage: the abaxial locule being fertile, the other sterile.

the point of leading to the union of the anthers of stamens of a single pair in the Gesneriaceae (the union takes place before anthesis).

Pistillodes (Sapindales)
Carpels could undergo reduction and transition to sterility, but that is less common than disappearance or abortion. Some carpels could indeed begin to develop and abort in favour of others, which is especially the case with the monocarpellary gynoecia of Laurales. The presence of pistillodes in some families is generally a derived character. For example, among the Sapindales (Rutaceae, Anacardiaceae, Sapindaceae, Meliaceae, Simaroubaceae, and Burseraceae), the Anacardiaceae (e.g., mango, *Mangifera* sp.) have pistillodes in the male flowers (these plants are generally dioecious), derived from the floral structure of Sapindales, of the type *, 4-5, 4-5, 4-10,(2-8). The conversion of fertile carpels into pistillodes is often a consequence of **floral sexualization** (and/or sexualization of the organism). In the same way, the female flowers of Anacardiaceae have staminodes. Thus, for the Anacardiaceae we write:

male: *, 5, 5, 5-10, (3-5)•

female: *, 5, 5, 5-10 •, (3-5)

3.4.2. Tiering

Till now, we have looked mainly at the architecture of flowers in which the receptacle is flat or forms a small dome, and the floral parts are thus attached practically at the same level. There are, however, flowers with derived structures on several levels (some parts are borne higher up) that are hereafter called **tiered** flowers. Flowers with a developed hypanthium are the first example of flowers with a more or less marked tiering (e.g., Rosaceae, Cactaceae, Aizoaceae). Two spectacular cases of tiered flowers are useful as illustrations: the Passifloraceae (e.g., passion flower) and the Zingiberaceae (e.g., ginger). These two cases differ greatly in the extent to which tiering occurs between the same whorls.

a) Passifloraceae (Malpighiales)

Often solitary, the actinomorphic flowers have 5 petaloid sepals that are free or very slightly connate, 5 free petals, an **outer corona**, an **inner corona**, 5 stamens, and 3 united carpels (parasyncarpous gynoecium), the 3 styles and stigmas being distinct (Fig. 3.23). The parts of the perianth are attached to a lax hypanthium, which has a **nectariferous disc** at its base. The outer and inner corona are filamentous **appendages of the hypanthium**, respectively long and short. Even though the hypanthium is not adnate to the carpels, the flower cannot be called perigynous because the stamens and carpels are borne considerably higher up on a sort of pillar, the

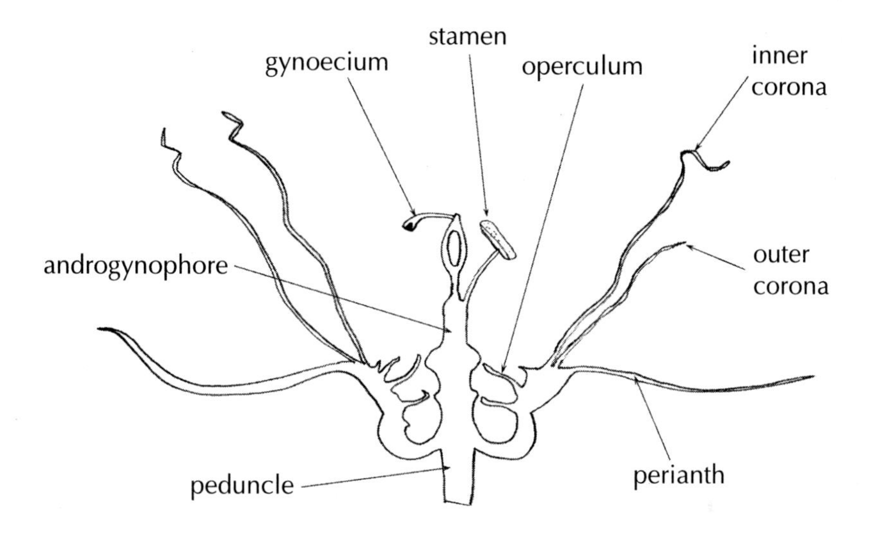

Fig. 3.23: General appearance of an axial section of a passion flower.

androgynophore.[13] The nectariferous disc is protected by a second appendage of the hypanthium, the **operculum**, the edge of which touches the androgynophore. However, not all the Passifloraceae have this architecture, particularly the androgynophore (*Adenia, Deidamia*). Passion flowers are pollinated by bees such as the Xylocopa, or even hummingbirds. The operculum plays an important role in this case, by preventing "cheats" from plundering the nectar.

b) Zingiberaceae (Zingiberales)

The Zingiberaceae have a derived floral architecture, considerably modified in relation to the ancestral architecture of type 3 of Commelinae (monophyletic group comprising notably the Arecales, Bromeliales, and Poales). The inflorescence is of the helicoid cyme type, camouflaged by large bracts. The flowers are zygomorphic, with 3 connate sepals, one of which is much larger than the others, 1 grooved stamen surrounding the style. There are also 4 staminodes, 2 of which are wide and united, forming a part of a type of **labellum** (a synapomorphy of the Zingiberaceae), and 2 others reduced, and 3 connate carpels, accompanied by a single style and a single stigma. The ovary is inferior. The calyx, reduced to a more or less tubular lobe, is separated from the stamen and the blade of the petals by a corolla **tube**, within which the

[13]Some Brassicaceae have a column supporting only the gynoecium, which is called simply the *gynophore*.

style is found. When the corolla tube emerges, the style is enveloped by the single stamen, so that it is never naked and only the stigma appears uncovered, above the anther (Fig. 3.24). Thus, the flower of Zingiberaceae is remarkable for the **shifting functions** of the flower parts: pollinators are attracted by the androecium (staminodes) and not by the petals, and the main part of the flower is carried by the corolla (tube) and not by the floral pedicel.

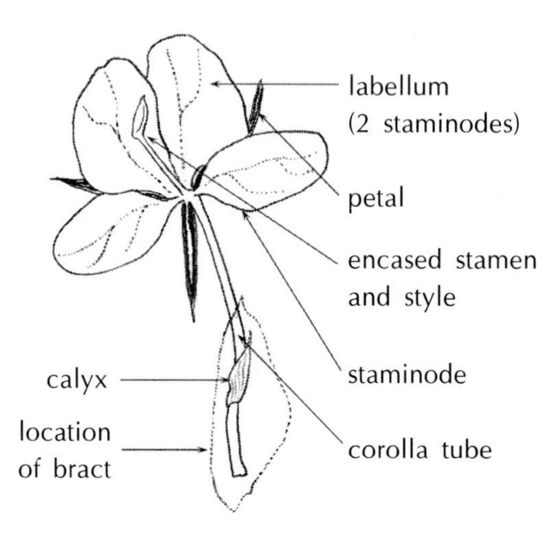

Fig. 3.24: Diagram of a detached flower of Zingiberaceae, external view.

3.4.3. Bracts and Modifications

a) General points

The bract is an element of the floral architecture. Some wrongly reserve the term bract to cases in which it is spectacularly differentiated. Still, such a distinction is misplaced. The bract constitutes a perfectly defined entity, appearing in some taxa in a derived form that is highly modified, supporting the entire floral structure. We have seen (in Chapter 1) its importance in the structure of inflorescences of Chloranthaceae. In the Tricolpates, it could replace the sepals (e.g., *Thunbergia*, Acanthaceae, see Fig. 3.19), or serve other functions. Among the Monocotyledons, the bracts are differentiated into **glumes** in the Poaceae, and into a cup or sheath in the Zingiberales. The bract indeed plays a highly significant role in the organization of this last group.

b) Bract, cup and sheath (Zingiberales)

Generally, the flowers of Zingiberales are enveloped in tough bracts, coloured or not coloured. In the Zingiberaceae, there is a **cup**, each containing one flower, most of which is at the tip of the fragile corolla tube. In the Strelitziaceae, such as bird of paradise, there is an inflorescence of the uniparous cyme type with 4 flowers enveloped in a bract of order 1, green and tough, and opening at the top, thus forming a **sheath**. The flowers have 3 orange sepals, 3 purplish petals, one of which (adaxial) is reduced and filiform, 5 stamens (the adaxial stamen has disappeared), and 3 united carpels. Not only does the inflorescence form a "boat" because of the bract, but also the flowers have a boat-like shape: the 2 abaxial petals are connate, sheltering the stamens (gigantic) and the style (Fig. 3.25). The flowers open and emerge from the bract one after the other. This architecture is not very different from that of the Fabaceae, in which the carina, made up of 2 united petals, shelters the stamens and style (under tension): this again is a phenomenon of **convergence**.

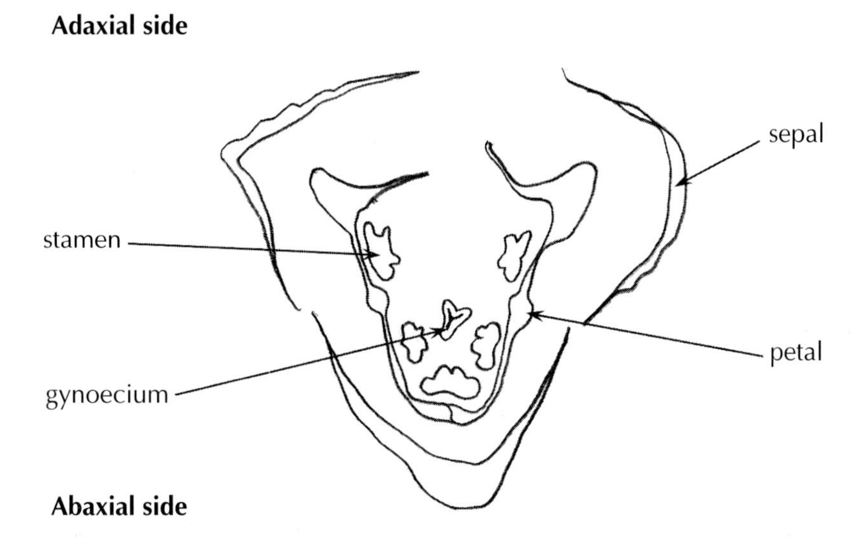

Fig. 3.25: General appearance of a flower of *Strelitzia* in cross-section. The adaxial petal does not appear because, being reduced, it does not emerge in the cross-section.

The entire flower is very solid and elastic, adapted to pollination by heavy organisms such as birds and small mammals (lemurs). The Heliconiaceae have a floral architecture similar to that of the Strelitziaceae; even though the inflorescences may be pendant or erect, the whole is not inverted, and the flowers face downward. In contrast, in the Marantaceae, the definite inflorescences have about 20 bracts, each enclosing 16 to 20 flowers. In the Cannaceae (monogeneric family: *Canna*), the bracts are membranous and soft and enclose 1 or 2 flowers.[14]

[14]The floral structure of Cannaceae and Marantaceae is not described in detail because it is highly complicated, especially with respect to the androecium.

The bract thus plays a very important role in the Zingiberales, which has repercussions for the floral structure: the enclosure of flowers in a single bract leads to flattening and a superimposed arrangement of flowers of a single cymule, and thus finally to the acquisition of lateral disposition of floral parts . The flowers of Zingiberales are profoundly zygomorphic. The Cannaceae or Marantaceae have only a single stamen and staminodes. The transition towards a staminodial state of stamens is quite characteristic of the Zingiberales.

4

Floral Architecture and Pollination

The subject of **pollination**, or the transport of pollen from the anthers to the stigma, can be discussed systematically, i.e., from one botanical group to the next, in order to show the wide range of possibilities that occur in the flowers of Angiosperms. We will not follow this approach, however, because our aim here is not to be exhaustive but to show that pollination (and fertilization) raises a series of biological problems, and that various strategies to overcome those problems have been observed.

Biological problems linked to pollination are not considered in the same way in autogamous and allogamous plants. Plants are called **allogamous** when a stigma of one plant is pollinated by a pollen grain from another plant. They are called **autogamous** when they are self-fertilized (whether by strict self-fertilization or geitonogamy, see Chapter 5).

In the case of allogamous Angiosperms, the major problem is that of **pollen transport**, which has three aspects: the output of pollen, the reception of pollen, and the transport agent. Moreover, Angiosperms presenting strict cross-fertilization are inherently susceptible to the deposit of self-pollen, and thus to **self-fertilization**. That means there are two problems: to favour cross-fertilization and to prevent self-fertilization. Autogamous plants are selected against, **to some extent**, for genetic reasons, because self-crosses lead to **inbreeding depression** (see Chapter 5).

4.1. DISPERSAL AND COLLECTION OF POLLEN IN ALLOGAMOUS PLANTS

4.1.1. General Characters

Pollination methods can be classified on the basis of **vectors**, i.e., the agents of pollen transport. There are plants with **biotic** vectors (insects, birds, etc.), called **zoophilous** or zoogamous, and those with **abiotic** vectors ("physical" agents such as wind or water). The biotic agents essentially comprise insects (**entomophily**), birds (ornithophily), bats (chiropterophily), and other small mammals. Bird and insect pollinators do not have the same shape or behaviour, and thus the constraints are different in each case. Interference between modes of pollination could cause reduction in the pollination efficiency, which is why there is often a close specificity between the flower and the pollen vector. Plants adapted to biotic pollen transport could be structured so that the wind and especially water cannot carry away the pollen grains. Just as a pendant flower plays the role of an umbrella, it has been proposed, for example, that the extended upper labium in some zygomorphic flowers functions to protect the anthers from raindrops and to keep the nectar from being diluted (Neal et al., 1998).

Before describing the structural details linked to pollination, we will give some general information about floral adaptation to different types of vectors.

a) Entomophily

Lepidoptera

There are three categories of Lepidoptera: the Noctuids, the Sphingids, and the Rhopalocera. Plants adapted to a mode of pollination via the Lepidoptera generally have elongated or even tubular corollas, in which the insect can easily insert its proboscis (maxillary *galea*). Sphinx moths are preferentially active at twilight or at night: the flowers that attract them are white or light-coloured, i.e., visible even in low light. The flowers produce nectar, of course, as well as perfume, according to a circadian rhythm, with a peak emission at night (e.g., *Stephanotis*, Asclepiadaceae). This type of flower may also bear a **spur**. This case has been a well-known example of floral biology since the studies of Darwin. Darwin supposed that there was a series of adaptations between the orchid and the insect. The flower tended to elongate its spur further and further so that the insect would hit its head against the stamens and stigmas, and the insect increasingly elongated its proboscis in order to reach the nectar easily. Thus, it was hypothesized that Orchidaceae with a long spur and insects with a long proboscis were selected for, leading to a considerable enlargement of these two structures, i.e., a process that favoured length. And that indeed seems to be the case: experiments tend to show that reduction of the length of the spur reduces the male and female fitness of orchids, following a less effective pollination.

Plants pollinated by Noctuids have a similar morphology, even though they have shorter tubes (the proboscis of Noctuids is generally shorter) and more vivid colours. Unlike plants pollinated by nocturnal moths, psychophilous plants (pollinated by Rhopalocera) bloom during the day, have short tubes, and offer a sort of "landing pad" for insects.

Diptera

Apart from the Hymenoptera, the Diptera are a very important group of pollinators: some also feed on pollen. It has been supposed that the Diptera were archaic pollinators, notably because the Magnoliidae are pollinated by flies. However, this hypothesis is yet to be confirmed. Flies are not highly specific; they visit many sorts of flowers and use nectar, stigmatal secretions, or even pollen. The flowers emit musky or even nauseating odours (amines such as cadaverine and spermine, with a putrid odour). These amines mimic the odours of cadavers or wastes, leading certain insect species to lay eggs on the flower; however, the larva does not develop and dies. From this point of view, the plant **parasitizes** the insect and kills the larvae for its own benefit. Similarly, some flowers mimic the host fungi on which larvae normally develop, in this case also by means of odours (*Asarum* sp., Aristolochiaceae). Some **myophilous** flowers (pollinated by flies), such as *Rafflesia* sp. (Rafflesiaceae), are among the largest flowers in the world.

Coleoptera

Cantharophilous flowers (pollinated by Coleoptera) produce large quantities of pollen or even special tissues devoted to the feeding of insects, in the stamens or the perianth. They also produce perfumes.

Hymenoptera

Pollinating Hymenoptera are mostly bumblebees, honeybees, or Xylocopa. Flowers pollinated by Xylocopa are generally more sturdy than those pollinated by smaller insects such as honeybees. They also have hidden sources of nectar that can only be reached by force. In temperate regions, the genus *Bombus* is a widespread group of pollinators. Flowers pollinated by insects of this genus generally produce nectar, except "deceptive" orchids (see Chapter 5). It has been supposed that flowers pollinated by bumblebees are often zygomorphic because of the insects' preference for bilateral symmetry. Even though bilateral symmetry may be associated to some extent with the possibility of formation of a lower lip allowing insects to land, and with the ability of insects to remember and learn (Neal et al., 1998), this hypothesis is still to be precisely formulated, particularly because experiments testing the symmetrical preferences of pollinators have yielded results that are quite variable from one species to another. For example, in fireweed (*Epilobium angustifolium*, Onagraceae), it has been demonstrated that pollinators (bumblebees) have a preference for symmetrical floral forms (Moller, 1998). Occasionally, flowers pollinated by other Hymenoptera could be visited by wasps, which use the nectar. Wasps generally have quite short mouth parts, and they generally seek nectar in flowers without a deep tube.

Other insects
Some other insects, such as thrips (Thysanoptera), visit flowers to look for pollen and possibly nectar. Flowers specialized in this mode of pollination are white, small, and narrow, and the stamens are prostrate.

b) Ornithophily

Pollinating birds include of course the hummingbirds as well as other groups such as the Psittaciforms (for example, some mimosas—*Acacia*—of Australia are pollinated by the rosehill parakeet). The ornithophilous flowers are brightly coloured (sight is a highly developed sense in birds), especially red (see Chapter 5), and fragrant. Hummingbirds have a long beak and a very long tongue, which allow them to drink nectar. The nectar rises by capillary action along the folded tongue, and the bird drinks (the bird does not suck as through a straw, contrary to a widespread notion). Ornithophilous flowers are consequently tubular, producing abundant and fluid nectar that makes it possible for the bird to feed. In addition, sclerified tissues are present around some organs, preventing the flower from being damaged by possible stabs of the beak. This mode of pollination is not exclusive, since in some cases the animal rests on the flower and transfers the pollen on its feet (*Strelitzia*, Strelitziaceae). Finally, there seems to be a correlation between epiphytism and ornithophily, especially in Gesneriaceae (Wiehler, 1983).

c) Chiropterophily

In a tropical environment, there are pollinating bats (pollinating Bignoniaceae, Fabaceae, etc.). These animals, active at night, are attracted by the fragrances emitted by the flowers (a process regulated by a circadian rhythm). The mucilaginous flowers produce a large quantity of pollen, which also feeds the bat. Frequently, these animals rest directly on the flowers to feed; thus, chiropterophilous flowers are large and the plants have for the most part an arborescent habit that gives them some solidity.

Some species are pollinated by mammals that do not fly. The Proteaceae and Myrtaceae of Australia are pollinated by small marsupials, and the Hypericaceae of Madagascar are pollinated by lemurs. More rarely, some flowers, particularly Melastomataceae of Costa Rica, are pollinated by rodents (Endress, 1994).

d) Hydrophily and anemophily

Many Angiosperms are pollinated by wind, an abiotic vector (**anemophily**). Generally, wind-pollinated plants do not have attractive floral parts and are made up essentially of fertile parts: stamens and/or carpels. Among the Malpighiales, in some Salicaceae, this type of pollen dispersal is a derived strategy, not an ancestral one. Anemophily has emerged independently several times in Angiosperms (e.g., Platanaceae, Fagaceae, Betulaceae). For example, in poplar (*Populus* sp.), the flowers are arranged in catkins. The wind carries the pollen, which is produced in large quantity. The inflorescences develop **before** the leaves are established, so that the action of the wind is not obstructed by leaves, and so that the pollen grains are easily received on the stigma. There are nonetheless Salicaceae in which the flowers, although they are arranged in catkins, have nectar and are pollinated partly by insects: for example, willows (*Salix* sp.). The entomophily of the genus *Salix*, within the Salicaceae, is a derived strategy. Anemophily exists in other families of arborescent Angiosperms such as Platanaceae (Proteales), as well as Fagaceae, Betulaceae, and Juglandaceae (Fagales), and all these families have wrongly been classed together precisely because of their anemophilous mode of pollination. In fact, the grouping and reduction of flowers constitutes an evolutionary **convergence** of the three groups (Fagales, Platanaceae, Salicaceae), directed towards pollen dispersal by wind.

Finally, a remarkable case of anemophilous pollen dispersal is that of the Monocotyledons Typhales, Juncales, and Poales (see box).

Some very unusual plants are pollinated by water (**hydrophily**). Notable examples are aquatic Angiosperms of the family Hydrocharitaceae (duckweed, *Vallisneria*). In *Vallisneria*, a dioecious plant, the female flowers remain attached by a long peduncle and float to the surface, while the male flowers detach themselves, reach the surface, and "navigate" like small boats. The pollen reaches the stigma of the female flower by direct contact or when the water is

Typhales, Poales and Juncales

These herbaceous plants have spike inflorescences, the flowers being unisexual or bisexual. In Typhaceae, the flowers are always unisexual, and flowers of the same sex are combined in dense spikes, the inflorescence in *Typha* having overall an upper staminal part and a lower carpellary part. The flowers have a membranous perianth sometimes reduced to barbules, and the filaments of stamens are more or less connate, so that the male flower is nearly reduced to anthers (Fig. 4.1). In *Sparganium*,* the stamens of male flowers have a relatively long filament, exposing the anthers to the wind (Fig. 4.1). Poaceae have pendant stamens with a long filament and a feathery stigma, favouring the emission and reception of pollen respectively (Fig. 2.27). In Cyperaceae, the bifid stigma and anthers extend far beyond the bract axillating the bisexual flower (Fig. 4.1). All these families produce pollen in large quantity (several hundreds or thousands of pollen grains per anther).

* This genus, earlier included among Sparganiaceae, is presently grouped with Typhaceae.

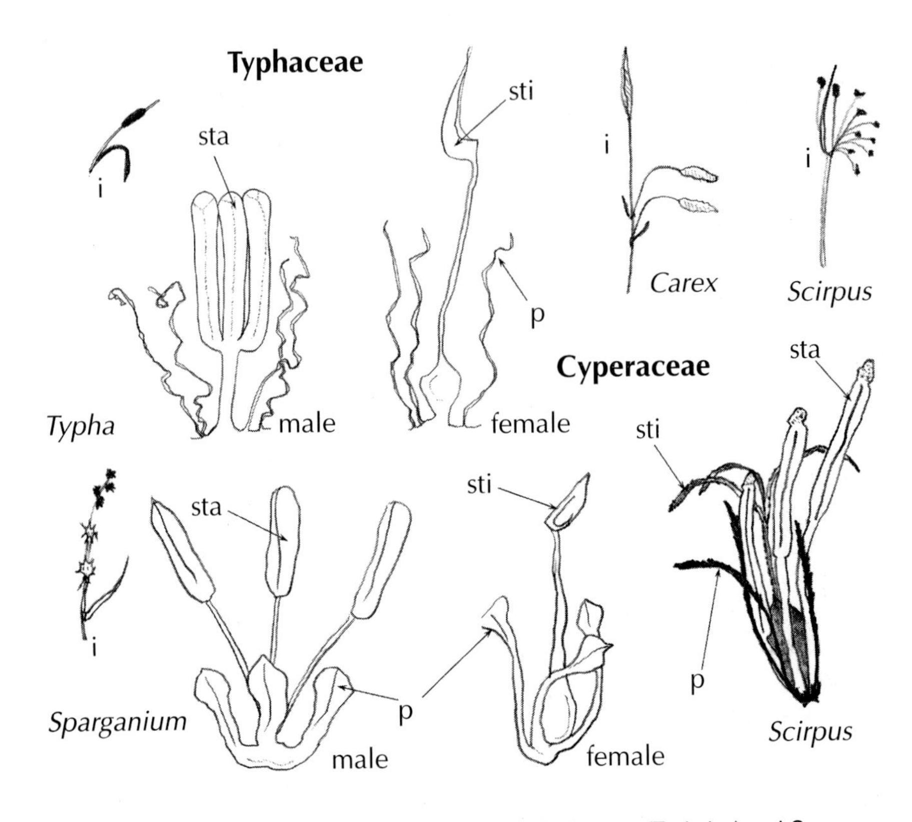

Fig. 4.1: Diagram of inflorescences (i) and flowers of Typhaceae (Typhales) and Cyperaceae (Juncales). sta, stamen; sti, stigma; p, perianth.

agitated. In duckweed (*Elodea canadensis*), on the other hand, the pollen drifts on the surface of the water until it reaches the filamentous stigma. Other Hydrocharitaceae, such as *Thalassia* sp., are pollinated entirely under water (the pollen is released in a mucilaginous secretion). Related families such as Posidoniaceae and Zosteraceae are also pollinated entirely under water. These three families, as well as Ruppiaceae, Najadaceae, and Zannichelliaceae,[1] have, among the Alismatales, derived modes of pollination adapted to aquatic life. It has been supposed that this mode of pollination was derived from anemophily, but that does not seem to be so: the other families of Alismatales are pollinated by insects (Alismataceae, Butomaceae, and Araceae), suggesting that the hydrophily arises from a specialization concomitant to the invasion of the aquatic environment, from an ancestor that was no doubt entomophilous (Fig. 4.2).

---

[1] And Potamogetonaceae, to a lesser extent.

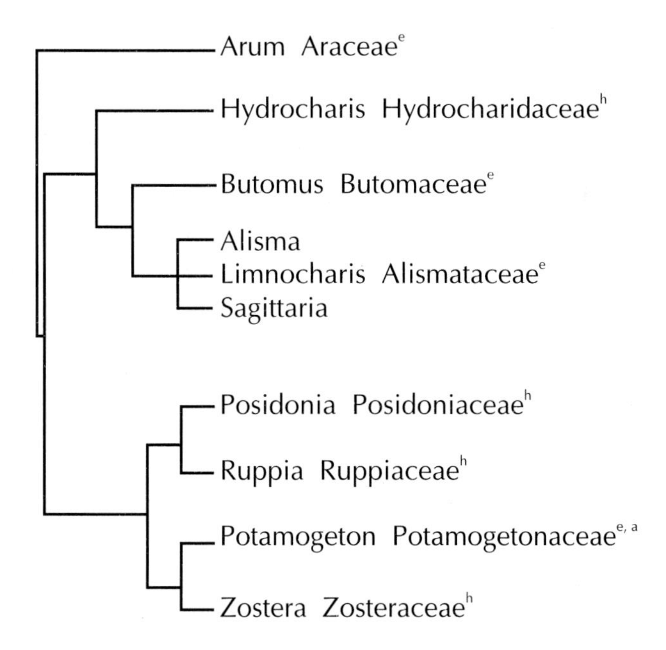

Fig. 4.2: Phylogenetic tree of Alismatales, based on the analysis of *rbcL* sequences. The modes of pollination of each family are indicated as superscripts: e, entomophily; h, hydrophily; a, anemophily. Potamogetonaceae have some wind-pollinated species, that is why the symbol *a* is added. (Modified from Bremer et al., 2000.)

4.1.2. Strategies to Attract the Vector

Pollen dispersal is dependent on pollen vectors in the zoophilous Angiosperms; the variety of floral structures and compositions are among other things intimately linked to strategies to attract pollinators. There are several levels of study in our discussion here: the population scale, the scale of the flower, and the molecular scale. As it is not possible to be exhaustive, at each step we will develop one or two demonstrative examples, avoiding a "catalogue" approach.

a) Group effect

Flowers are generally not isolated because, first, they are carried by an inflorescence and, second, they belong to a plant that is itself surrounded by congeners of the same or other species.

Inflorescences

As we have already mentioned (see Chapter 1), flowers are grouped into more or less compact inflorescences. A large number of flowers tends to increase the attraction exerted on pollinators, because the emission of perfumes is more intense, the flowers are more visible, and so on. Moreover, the grouping of flowers into an inflorescence offers the possibility of **floral differentiation**. This is particularly visible in Asteraceae (see Chapter 2, section 2.2.1), subfamily Asteroidae. The outer flowers are differentiated into ligulate ones. This has important repercussions for the floral structure: the outer flowers are generally partly sterile and zygomorphic, unlike the other flowers of the inflorescence. Moreover, they sometimes do not have the same colour as the others: white in the case of marguerite (daisy), mauve in aster, etc. The sepals are always highly modified, being reduced to the state of filamentous, feathery, or silky parts forming a **pappus**. The differentiation may also imply a reduction of all the flowers in favour of the bract, which takes on the role of attraction. For example, in lavender (*Lavandula stoechas*, Lamiaceae), the flowers at the tip of the cluster have violet bracts in the form of a feather, easily visible to pollinators (the destruction of these bracts leads to a considerable drop in the number of visits by bumblebees or honeybees (Fig. 4.3)).

Homospecific groups

The juxtaposition of plants of the same species in some cases considerably increases the number of visits by pollinators, for the same reasons given above. For example, in lavender, the number of visits is very low when the populations are scattered. Conversely, the denser populations have a higher rate of visits and produce statistically larger seeds, probably favouring the eventual germination because of a greater quantity of reserves. This difference probably comes from the fact that the scattered populations are susceptible to

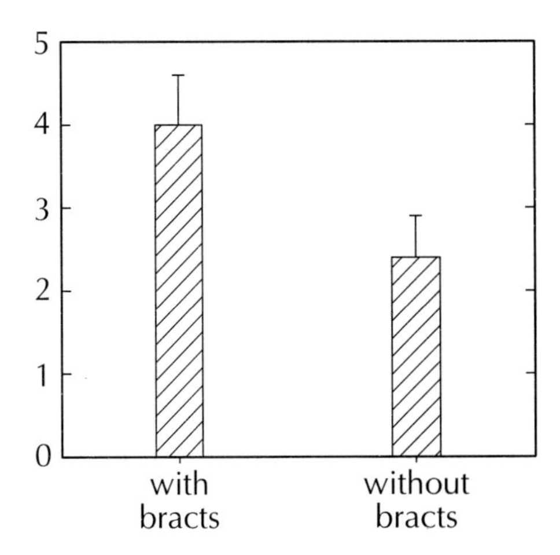

Fig. 4.3: Average number of visits by pollinators (bees) over 30 minutes on lavender plants (*Lavandula stoechas*) with or without apical bracts. In this experiment, a pair of lavender plants was isolated, the bracts removed from one of them, and the choice made by the pollinator observed. In these conditions, the insects prefer the individuals with bracts. (Modified from Herrera, 1997.)

geitonogamous self-fertilization, unlike denser populations; in a scattered population, the pollinator more often chooses to stay on a flower of the same plant. In some plants, vegetative multiplication results in tufts, which locally increases the number of flowers and also has a similar effect; this is true for lupin (Fabaceae).

Heterospecific groups

In the same way, plants of different species could be juxtaposed and increase the attraction exerted on pollinators. The attraction works better if there is a **resemblance** between the two species. Without necessarily implying juxtaposition or massive groups, some species could benefit from the attraction exerted by a similar-looking species. This shows that **mimicry** is not rare among the Angiosperms. The mimic species exploits the pollinator's learned recognition of the mimicked species and thenceforth is abundantly visited, without necessarily offering anything in return (nectar). The mimicry is, from this point of view, positive for the mimic species, increasing its rate of visits and its fitness.

There are two types of mimicry: **Batesian** and **Mullerian**. These two types are distinguished by the dynamics of the abundance of the mimic species (Fig. 4.4). In the Batesian case (e.g., *Orchis israelitica*, Orchidaceae, and *Bellevalia*

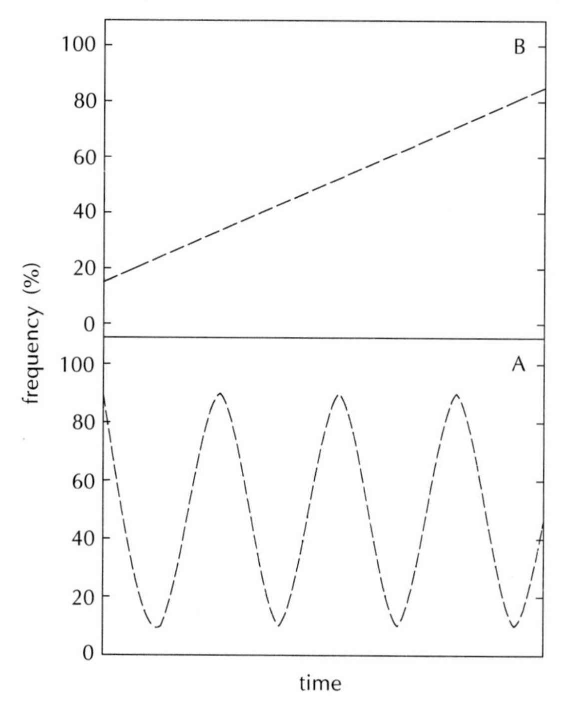

Fig. 4.4: Appearance of the evolution of the frequency of the mimic species as a function of time in Batesian mimicry (A) and Mullerian mimicry (B) (modified from Roy et al., 1999).

flexuosa, Liliaceae), the mimic species is increasingly abundant at first, benefiting from the pollinators of the mimicked species. Then, the "deceived" pollinators learn to associate the mimic flower and the lack of nectar and ultimately avoid the mimic flower. The pollination declines, the mimic species becomes less abundant. Finally, when the mimicked species becomes more frequent, the pollinator again learns to associate the mimicked flower and the supply of nectar. It happens to visit the mimic flower again, and the cycle renews itself. Thus, in this case abundance has a **negative effect** on the increase of the mimic population. In Mullerian mimicry (e.g., *Senecio integerrimus* and *Helenium hoopesii*, Asteraceae), the opposite occurs: the rate of increase depends positively on the frequency of the mimic. In Mullerian mimicry, two or more species find an adaptive advantage in flowering together in dense populations, increasing the probability of visits by pollinators. They have one or more common characteristics, resulting from evolutionary **convergence**. These two cases of mimicry are thus very different: Batesian mimicry never offers a benefit (nectar) to the pollinator, which is not necessarily so in cases of Mullerian mimicry. Moreover, the mimicked and mimic species are clearly distinguished in the Batesian case, whereas it is not inherently clear in the Mullerian case which is the "model" species.

Mimicry has, however, a potential disadvantage, which is interspecific pollen transfer. In the Orchidaceae, where there are a large number of mimic species, pollinia fit correctly only in flowers of the same species. In other species (Zingiberaceae), the flowers persist until homospecific pollen grains are deposited on the stigma (Roy et al., 1999). In general, however, it is probable that although selection can in some cases favour similarity, there is also a selection pressure favouring the emergence of differences that reduce the rate of what can be called "improper" pollen transfer.

b) Modifications

"Modifications" refer to the acquisition of derived floral structures, specialized for one particular type of pollination. Three examples are discussed in this section: Orchidaceae, Lamiales, and Ranunculales, which are excellent cases of specialization-divergence from actinomorphic type 3, zygomorphic type 5, and helical actinomorphic ancestors respectively.

Orchidaceae

Orchidaceae, which belong to the order Asparagales, have a zygomorphic structure (see Chapter 3) of type 3. The perianth, the most visible part, is composed of 3 petaloid sepals (tepals) and 3 petals in which the abaxial[2] is differentiated into a **labellum**. Apart from this mark of zygomorphy, the floral structure is highly complex. The stamens do not have a typical filament-anther

[2]Note the inversion of the adaxial-abaxial polarity in Orchidaceae; see Chapter 3.

structure but are reduced to locules into which the pollen, aggregated into pollinia, are inserted. The pollinia comprise a sticky base (the retinaculum), a stalk, and a body. The pollinia are opposite to the abaxial petal, which suggests that they are derived from the adaxial stamen, the others remaining undeveloped. The gynoecium comprises 3 united carpels with parietal placentation. The stigma, highly modified, presents a non-receptive part made up of a rostellum, which creates an "obstacle" between the stamen and the receptive zone of the stigma, and a viscidium, sticky and attached to the pollinia. The flower also has characters that are the source of specialization with respect to a pollinator or group of pollinators: ornamentation of the labellum that imitates the body of the insect (*Ophrys* sp.), emission of fragrances imitating the sexual pheromones of the insect (*Ophrys ryptostylis*), and correspondence of the flower size with that of the insect. Also, some Orchidaceae produce nectar and others do not (*Cypripedium*).

This descriptive outline applies particularly to Orchidoidae. In Apostasioidae, for example, the stamens, numbering 3, are almost normal and slightly adnate to the gynoecium (Fig. 4.5). The gynoecium mostly has axile and not parietal placentation. Finally, there are no pollinia and the pollen is present in the "classic" form of individual grains. The other sub-families (Orchidoidae,

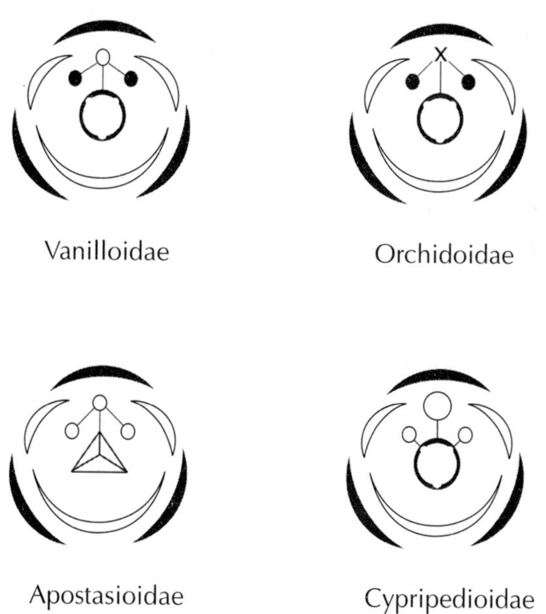

Fig. 4.5: Floral diagrams of sub-families of Orchidaceae. Apostasioidae have a eusyncarpous ovary (axile placentation), while the other sub-families have parasyncarpous ovaries (parietal placentation). In Cypripedioidae, the upper stamen(s) is modified into staminode(s). In monandric Orchidaceae (Vanilloidae and Orchidoidae), there are pollinia. The upper stamen is functional (Vanilloidae) or not functional (Orchidoidae). Sometimes, the lower sepals are united in Cypripedioidae (not illustrated).

Cypripedioidae, Vanilloidae) have an agglomerated and sticky pollen, as well as an asymmetrical stigma. In Cypripedioidae, the adaxial stamen is transformed into a staminode and there is no pollinium. Vanilloidae, finally, have a single functional stamen (the adaxial) and sometimes have no pollinia. Thus, the monandric Orchidaceae (Vanilloidae and Orchidoidae),[3] a probably monophyletic group (Freudenstein et al., 1999), are distinguished from the **polyandric** Orchidaceae (Apostasioidae and Cypripedioidae), which are undoubtedly paraphyletic.

Generally, pollinators attracted by the flowers of Orchidaceae rest on the labellum and penetrate the flower in search of nectar, a process during which the pollinia stick to their heads or other body parts. When the pollinator goes to the next flower, the pollinia on its body strike the stigma of that flower, which leads to fertilization. The visual or olfactory floral traits certainly make for a high **plant-pollinator specificity** but do not guarantee the absence of self-fertilization in the wider sense. Indeed, some Orchidaceae are susceptible to geitonogamous self-fertilization (see Chapter 5) or even strict self-fertilization, for example, vanilla. In *Vanilla planifolia*, a small dipteran (*Melipona*) visits the flower and triggers self-pollination when it brushes against the rostellum.

This highly modified, derived organization, among the Asparagales and Orchidaceae, is supplemented by other characters related to the floral specialization that we have referred to (see box).

These organisms thus have varying floral architecture and more generally different floral traits adapted to various modes of pollination. Even though the three species described in the box are quite closely related phylogenetically, they demonstrate the spectacular adaptive range of Orchidaceae.

Examples of Floral Specialization of Orchidaceae

Let us take three different examples of Orchidoidae found in the temperate regions of the northern hemisphere: *Ophrys holoserica, Anacamptis pyramidalis*, and *Listera cordata*. In *Ophrys*, the sepals are reduced, the labellum is decorated with plumules and a bright and dark motif, the **sperculum**, which makes it look like an insect of the Apidae type. The flower also gives off an odour. All these characteristics attract male bees (*Eucera*). In *Anacamptis*, the almost uniformly pink flower has a long spur in which nectar is produced and in which circadian **butterflies** can insert the proboscis. The flower of *Listera cordata* is of a dull greenish or purplish colour; the labellum is clearly bifid. The pollen is released explosively in a sticky suspension. The flower gives off an odour of putrefaction. This attracts Diptera (flies), which rest on the flower and end up with pollen sticking to them.

[3]Sometimes, the monandric Orchidaceae are divided into three sub-families: Vanilloidae, Orchidoidae, and Epidendroidae (e.g., *Epidendrum*), the last group comprising mostly epiphytic orchids.

Lamiales

Like Orchidaceae, Lamiales are a group including species that, even though they are all constructed according to a zygomorphic type 5 architecture, have structural differences, indications of adaptations to varied modes of pollination. The different families that make up the order Lamiales have already been cited in Chapter 3.

The flowers of Lamiales have connate petals (gamopetalous corolla); consequently, the flower frequently has a tubular shape. However, while the tube is short or wide in Lamiaceae, Plantaginaceae, and other groups, it is long and thin in *Buddleia*. Moreover, this genus does not show zygomorphy in that it does not have a dilated lower lip. In other words, *Buddleia* is peculiar to the extent that the flower, lacking a "landing pad" and forming a slender tube, is adapted to pollination by **butterflies**. The flower also has a fragrance that attracts these insects. *Buddleia* is therefore sometimes called the butterfly tree. The flowers of Lamiaceae, on the other hand, have a large lower lip, a marked bilateral symmetry, and ornamentations or trichomes that serve to guide pollinators. Some species also have a mechanism for pollen deposit consisting of stamens with "pedals", a typical example of which is sage (*Salvia* sp.). Indeed, when the insect enters the flower, it rests on the proximal part of the adaxial stamens, causing the lowering of the distal part and thus the anthers. By contrast, Orobanchaceae are pollinated most often by Diptera; the flowers have little colour or brightness and have a putrid odour.

Finally, some species among Gesneriaceae or Lamiaceae are ornithophilous and have a very slender tubular flower structure, perfectly adapted to the insertion of the beaks of pollinating hummingbirds.

Ranunculaceae

The variety of the Ranunculales, rarely studied, is remarkable (Fig. 4.6). Ranunculaceae present actinomorphic flowers (*Ranunculus*) as well as zygomorphic flowers (*Aconitum*). While the yellow actinomorphic flowers are not specialized in terms of pollination mode, the purple (or yellow) zygomorphic flowers mostly attract Hymenoptera. Some genera have nectaries at the petals (sometimes thinning into a spur or dilated into a helmet—*Aconitum*), producing large quantities of nectar (*Delphinium, Aquilegia*), attracting mainly bees or bumblebees. In *Caltha*, the nectaries are at the base of the carpels. *Anemona* and *Clematis*, on the other hand, do not produce nectar and attract insects that harvest the pollen. At the opposite end, some Ranunculaceae, such as *Thalictrum*, are anemophilous: the parts of the perianth fall rapidly, the stamens are numerous and gaudy, and pollen is effectively dispersed by wind.

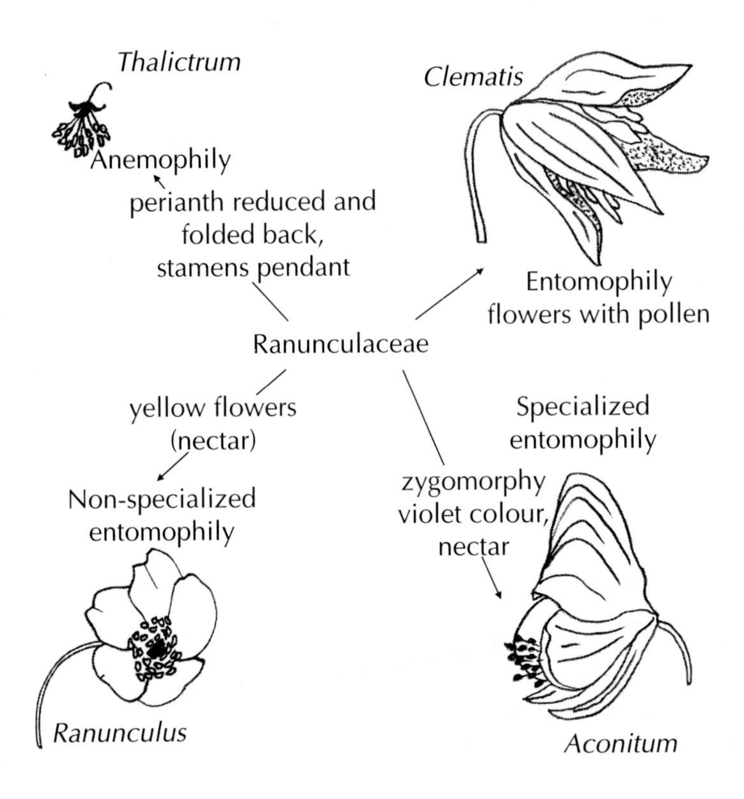

Fig. 4.6: Some examples of modes of pollination in Ranunculaceae. In each case, a genus is cited as an example (see text). This diagram is not a phylogeny.

c) Nectar

General points

Many flowers produce nectar, a sweet liquid on which pollinators feed. Nectar is produced in specialized organs, the floral **nectaries**. Flowers are not the only parts of the plant to produce nectar. In some species, as in Passifloraceae, there are what are called "extrafloral" nectaries. Nectar is a sort of "reward" for the pollinator, which ultimately associates the flower (of that particular species) with the presence of nectar. It is made up of water, mineral salts, amino acids, saccharose, and hexoses such as fructose and glucose. The amino acids are present in low quantities, around 1 millionth the quantity of hexoses. Nectar also contains vitamins, alkaloids, phenolic compounds, antioxidants, and saponins in small quantities.

Nectaries may be found in different parts of the flower: sepals, petals, stamens, carpels. The most common are **nectariferous discs**, located between the stamens and carpels, in Dicotyledons (Fig. 4.7), and **septal** nectaries, located in the septa of the ovary, in Monocotyledons (Endress, 1994).

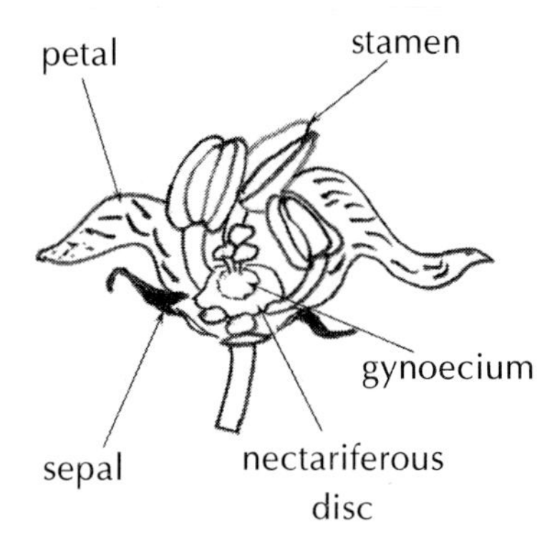

Fig. 4.7: Diagram of a flower of Anacardiaceae showing a nectariferous disc at the base of stamens. Two stamens and part of the perianth have been removed in order to reveal the inside of the flower.

Even within a single flower, there may be several types of nectaries. For example, in Brassicaceae, there are **lateral** nectaries (associated with 2 short, lateral stamens) and **median** nectaries (associated with long stamens). Species of the family Brassicaceae may have both types of nectary or a single type. The two types differ in their sugar composition: the lateral nectaries have a glucose/fructose ratio of about 1.1, while the median nectaries have a ratio of 0.2 to 0.9. Moreover, lateral nectaries have a significant and profound phloem vascularization, which is not the case with median nectaries (Davis et al., 1998). The biological significance of this heteromorphy of nectaries is still to be understood.

Typology and modalities of production

Nectar is produced directly by the epidermis (**epidermal** nectaries), the mesophyll (**mesophyll** nectaries), or even specialized trichomes (**trichome** nectaries). Mesophyll and epidermal nectaries have stomata at their surfaces, in which the guard cells have lost the capacity to close the ostiole. In other words, these stomata do not regulate nectar flow. Whereas epidermal and mesophyll nectaries are found often, including among the extrafloral nectaries, trichome nectaries are more rare (e.g., Malvaceae, Dipsacaceae). The phloem irrigates the nectaries, providing organic molecules that are used in nectar production, particularly photosynthates. The cells that produce nectar have **starch**, which is probably a result of the polymerization of glucose from the conversion of imported saccharose. This starch is subsequently used to produce glucose, fructose, or saccharose, which accumulate in the nectar.

Nectar was earlier thought to be secreted by means of an active transport of hexoses or saccharose by a membrane transporter (Luttge, 1969), similar to the apoplastic load of phloem assimilates. However, it seems clear that nectar is secreted by exocytosis. The exocytotic vesicles come directly from the cisternae of the endoplasmic reticulum (Ed Echeverria, 2000). Still, it is possible that part of the structures supposed to be ER in the microscopic preparations are a result of an artefactual fragmentation of the vacuole during fixation (Verbelen et al., 1998). In any case, the implication of the vacuole in nectar synthesis is presently uncertain; it has been suggested that vesicles of ER and vacuolar origin could fuse, effectively forming exocytotic vesicles. Finally, nectar-producing cells have invaginations of their plasma membrane, a sign of their significant exocytotic activity (Fahn et al., 1970).

Nectar and pollination

Nectar composition varies greatly according to the mode of pollination. Flowers pollinated by hummingbirds, butterflies, and Diptera having an elongated labium have a high saccharose/hexose ratio, while those pollinated by flies, perching birds, or bats have a low saccharose/hexose ratio. The preferences of Hymenoptera such as bees are not clearly defined at present. On the other hand, these insects prefer high concentrations of sugars, which is not true of hummingbirds or butterflies. Hummingbirds and butterflies probably use the least viscous nectar, which is better adapted to their mode of drinking, i.e., by means of a fine conduit (respectively the tongue and the proboscis).

That said, there is a wide **variability** in nectar composition from one flower to another in an inflorescence or from one plant to another, and even from one time to another. Nectar production has been particularly studied in the genus *Epilobium* (Onagraceae). In *E. canum*, nectar production varies from 6 to 117% from one flower to another. The environment also influences nectar production, in that drought or low light induces a significant drop in production (Boose, 1997). In *E. angustifolium*, a prolonged dry spell (12 days) causes a drop in nectar production but does not induce a change in the sugar concentration (Carroll et al., 2000). Finally, nectar production in this same species depends on the floral architecture: asymmetrical flowers produce significantly less nectar than symmetrical ones. Floral symmetry is thus perceived by pollinators as an indication of the available quantity of nectar (Moller, 1995).

Finally, in rare cases, once pollination occurs, the remaining nectar is reabsorbed to some extent. For example, in orchids of the genus *Aerangis*, pollination induces a reabsorption of around 70% of soluble sugars (Koopowitz et al., 1998). It is probable that the reabsorption of sugary compounds of nectar is a means of limiting the energy cost of nectar production.

Development of nectaries

Floral nectaries are an integral part of the flower, which raises the question of the determination of their development in relation to the "ABC" genes. The development of nectaries has especially been studied genetically in *Arabidopsis*, a species with two types of nectaries (median and lateral, see above) located at the base of the stamens. First of all, it has been observed that the mutations *ap-1, ap-3, pi,* and *ag* do not result in the disappearance of nectaries. On the other hand, the double mutations *pi ag* or *ap-3 ag* result in the formation of an entirely sepaloid flower that lacks nectaries, suggesting that the disappearance of whorl 3 (where the stamens are located in the wild type) leads to the disappearance of the nectaries. In the mutant *ufo*, in which the function B is no longer activated, the flower does have nectaries in whorl 3, which is composed of carpels. This suggests that nectaries form in whorl 3 (when it exists), independently of the organs that develop in that whorl.

Furthermore, the mutation *crc* (see Chapter 2) leads to the disappearance of nectaries, while the overexpression of *ufo* (obtained by promoter 35S) leads to the formation of supplementary nectaries in whorl 4, as does mutation *sup* (the gene *sup* seems to inhibit the expression of *pi* and *ap-3* in whorl 4) (Baum et al., 2001). Still, the overexpression of *pi* or *ap-3* does not lead to the formation of nectaries in whorl 4 (which could be explained by the inhibition exerted by *sup* as well as the absence of the expression of *ufo* in this whorl). It can thus be supposed that the genes "ABC", *ufo*, and *sup* are responsible for the position of nectaries, while the induction of their development strictly speaking depends on other genes such as *crc* as well as *ufo* and *sup* (see Fig. 4.8). The independence of nectaries with respect to the nature of organs with which they are associated (stamens and carpels) in *Arabidopsis* agrees with the observation that nectaries develop on various organs in different species, including leaves (extrafloral nectaries).

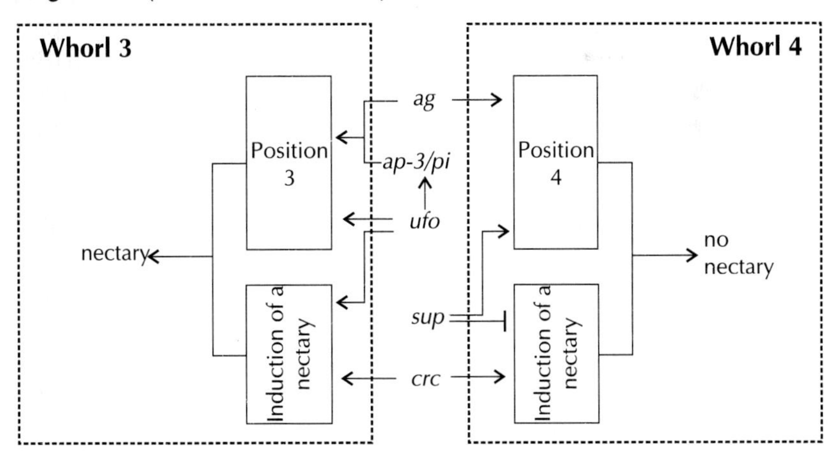

Fig. 4.8. Determination of nectary development in *Arabidopsis thaliana*. The positional information (whorl) is given by the expression of genes *ap-3, pi,* and *ag,* and also probably *ufo* (which stimulates *ap-3* and *pi*). The nectary development is induced by the expression of *crc* and *ufo* and inhibited by *sup*. When the conditions (position, induction) are combined, the nectaries develop. This diagram is simplified to the extent that there are certainly other genes that are presently unknown and that are linked to the development of nectaries. Arrows, activation; bar, inhibition.

d) Other resources offered by flowers

Apart from nectar, flowers produce other resources used by insects: **pollen** and **oils**.

Pollen

Flowers that offer their pollen as a resource do not produce nectar and are generally pollinated by Coleoptera or Apidae. Pollen has reserves such as starch or lipids that are used ultimately during pollen germination. Thus, pollen grains are a useful resource for insects in terms of nutrients. Pollen is produced by a small number of stamens with mostly poricidal dehiscence (e.g., Solanaceae) or by a large number of stamens (e.g., Papaveraceae, Magnoliaceae). In the latter case, the pollen could be light and powdery (Papaveraceae) or quite sticky (Magnoliaceae). The flowers that have few stamens are sometimes upside down (bittersweet or *Solanum dulcamara*, Solanaceae) and vibrations caused by the insect cause the pollen to fall through the anther pores. These flowers have large and showy anthers, short filaments, and often connate stamens, revealing the androecium (Fig. 4.9). In some species, there is a staminal heteromorphy: small and showy stamens produce the pollen effectively used by the insects, and larger but more discreet stamens produce pollen that is meant to cover the insect's body (e.g., Lecythidaceae). The pollen that is meant to be consumed is most often sterile. The fall of pollen is "channelled" by expansions of the corolla, which faces the anthers (e.g., Sapotaceae, Fig. 4.10) or by a **hood** (or cucullus) originating from the stamens (e.g., Commelinaceae) or corolla (e.g., Fabaceae Cesalpinioidae). The hood more or less squeezes the anthers, guiding the trajectory of the pollen and limiting the quantity released during each visit.

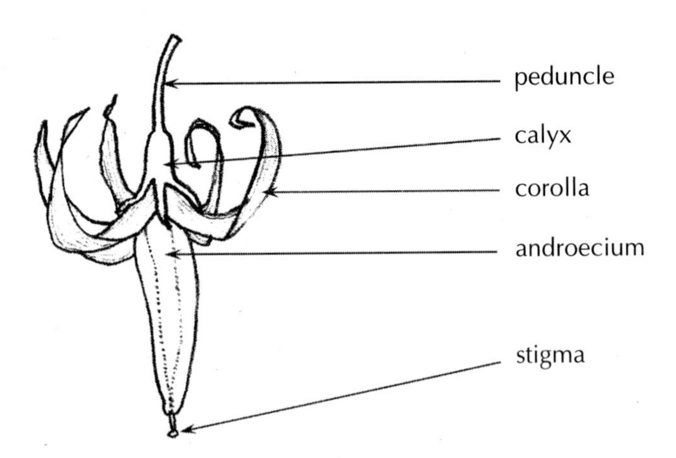

Fig. 4.9. Example of "pollen flower", bittersweet (Solanaceae).

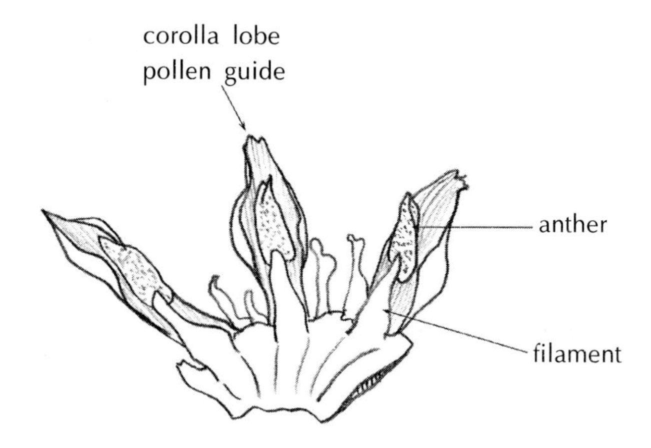

Fig. 4.10. Diagram of a portion of the flower of Sapotaceae showing corolla lobes that serve as a "pollen guide".

Oils

Some flowers, mostly tropical ones, offer oils to specialized pollinators (Hymenoptera), which use these compounds to feed their larvae (Vogel, 1979, 1988). The oils are produced by specialized glands, the **elaiophores**. They are collected not by the mouth parts of insects but by the feet or abdomen. There are **epidermal** elaiophores and **trichome** elaiophores (e.g., *Calceolaria*). In the former (e.g., Malpighiaceae, Orchidaceae), the oil accumulates under the cuticle and oozes through a thin part of it. In the latter (e.g., Iridaceae, Primulaceae), the oil-producing cell (which is the single cell of the trichome or the terminal cell of multicellular trichomes) forms a virtual vial containing the oil. The oils are compounds of mono- or diglycerides, the fatty acids (like C18:1) being frequently hydroxylated.

e) Colours

General points

The colour of flowers is involved in the optical attraction of pollinators. The corresponding wavelengths belong to a visible or invisible spectrum; some pigments emit light in the ultraviolet part of the spectrum, to which honeybees are sensitive. Mammalian photoreceptors are of the blue, green, and red type (similar to those in some birds), which is not the case with insects. Red, for example, is not detected by Apidae, but birds see it. Yellow flowers (e.g., buttercup) are visited by a wide range of insects and characterize non-specialized plants. In contrast, flowers pollinated mainly by honeybees or bumblebees are blue (e.g., sage) or purplish (sometimes pink). Flowers pollinated by hummingbirds are often red (see also Chapter 5). Flowers blooming at night, finally, are usually light coloured or white.

Pigments

There is a wide variety of floral pigments in biochemical terms (Table 4.1). There are **flavonoids** (comprising flavones, anthocyanins, chalcones, and aurones), betalains, and carotenoids (comprising carotenes and xanthophylls). Each class of pigments has a different biosynthesis pathway and cellular location. Flavonoids are generally in the vacuole, while carotenoids are in the plastids. Pigments are not simply stored as such in the cell. Anthocyanins, and even the majority of flavonoids, are present in solution in the vacuole, **combined** with sugars. In morning glory (*Ipomea tricolor*), the whole is made up of one molecule of peonidin, 6 molecules of glucose, and 3 molecules of caffeic acid (Fig. 4.11). The flavones (and flavonols, which differ from flavones only by C_3 oxygenation) are very often combined with a hexose (i.e., glycosylated) such as glucose or rhamnose. The betalains are also associated with oligosaccharides and other compounds, such as DOPA in bougainvilein. The carotenoids are present in the vesicles or tubules. The vesicular (or tubular) phospholipids and carotenoids are stabilized by particular proteins, such as Chrc, isolated from the corolla of the cucumber (Cucurbitaceae). The synthesis of these proteins is coordinated with that of carotenoids, because the inhibition of biosynthesis of carotenoids with norflurazon causes a decline in the transla-

Table 4.1. Floral pigments

Name	Flavones	Anthocyanins	Betalains	Chalcones, aurones	Carotenes	Xanthophylls
Colour	White to yellow	Red to blue	Pink-red	Yellow	Yellow to orange	Yellow
Example	Luteoline	Peonidin	Bougainvillein	Butine	Lycopene	Violaxanthine

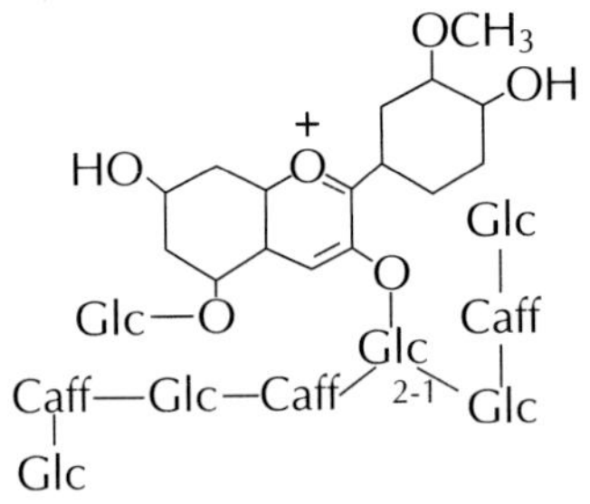

Fig. 4.11: Chemical structure of pigment of blue petals in morning glory (*Ipomea tricolor*). The pigment consists of 1 peonidin molecule (anthocyanin) linked to 6 glucose molecules (Glc) and 3 molecules of caffeic acid (Caff). The cafeate-glucose linkages are ester linkages attached to 6 glucose and the glucose-cafeate linkages are ether linkages attached to 1 glucose. The glucose-glucose linkage is of type 2-1. The whole molecule is located in a vacuole.

tion of RNA coding these proteins (Vishnevetsky et al., 1999). Chrc production is also regulated by hormones: gibberellins and ethylene activate the synthesis of pigments and of this protein, while abscisic acid has the opposite effect.

Some botanical groups produce compounds specific to them. The chemical compositions of plants can be used to establish a classification (**chemotaxonomy**). This is the case with some pigments: whereas the anthocyanins are widely represented in the Angiosperms, betalains are characteristic of Caryophyllales.

The colours of some flowers result from a **structural effect**. The existence of abundant intercellular spaces in the mesophyll diffuses light, making it appear white (Fig. 4.12). Most white flowers are of this kind. More rarely, some flavones give the flowers a white colour. This structural effect exists also in coloured flowers. The pigments are generally contained in the epidermis, and the intercellular spaces below the epidermis (in a layer called **optic tapetum**) make the colour brighter. In contrast, there is no colour, properly speaking (other than white), that is of purely structural origin, as is observed in the birds (throat feathers of pigeon, tail of peacock).

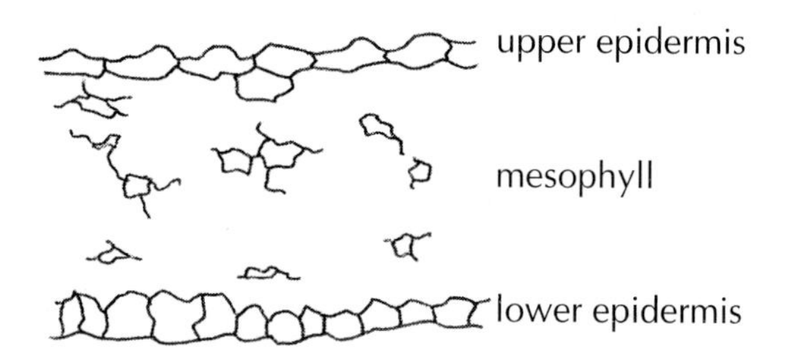

Fig. 4.12. Cross-section of a petal with a mesophyll rich in intercellular spaces that serve as an optic tap (m). If the epidermal cells lack pigments, the petal reflects light effectively and appears white. When the upper epidermis has a pigment, the optic tap makes the colour more vivid. Only the cell walls are visible in this figure.

In some flowers, the pigments are juxtaposed (adjacent cells do not have the same colour) or superimposed (there are several pigment layers). In the former case, the floral colour is said to be **additive** and in the latter it is called **subtractive**.

Colours and attraction

The typology of modes of pollination has often been placed in the context of a pollinator's vision of colours, so that it was considered possible to use the colour observed as an unambiguous indicator of the type of insects frequenting the flower. But things are not so simple: there are colours that insects see and colours they do not see. In other words, insects do not "see" flowers in the same way we do. First of all, there are three types of photoreceptors in insects:

a blue receptor, a green receptor, and an ultraviolet receptor (the wavelengths change slightly from one species to another). Bees see colours from 360 to 530 nm. From that observation to the statement that insects do not see poppies (*Papaver rhoeas*, Papaveraceae) is a short step, that has often been taken. In fact, some insects see wavelengths in red, such as butterflies, wasps, and Coleoptera; other insects, which do not have this faculty, nevertheless see a "patch" in a given field, either by contrast or because red flowers reflect enough blue light for insects that have a chromatic sensitivity in the green-yellow (around 540 nm). Besides, it is supposed today that the effects of borders and contrasts (**receptor field** theory) between two coloured zones contribute greatly to the detection of flowers and their shapes by insects (Horridge, 1998). For example, a red flower on a green background appears as a black patch on a coloured background, and bright yellow flowers appear as a brilliant patch on a dark background. That means that insects do not "see" flowers as we do, that is, that they do not draw the same information that we do: while humans dwell on the nature of the colour, insects look at other information about the floral morphology, such as contrast, edges, shape, and symmetry.

Finally, many floral traits may not be independent, and especially colour, which is found to be linked, in snapdragon, to the rugosity of the epidermis. Indeed, the *mixta* mutants of snapdragon are visited less often than the wild type; in this mutant, the flat and non-conical surface of the epidermal cells gives it a light pink colour, unlike the wild type, which is magenta (Waser and Chittka, 1998). In natural populations of snapdragon, the light pink flowers seem to produce less nectar, which explains the preference of pollinators for the magenta flowers. Moreover, the difference in epidermal rugosity, which insects perceive by their tarsal receptors, could affect the preference of pollinators. The albino flowers of snapdragon (*nivea*) are also visited less frequently than those of the wild type, undoubtedly because the corolla reflects ultraviolet rays in these mutants, which makes the flowers appear colourless to insects.

f) Perfumes

The perfumes emitted by flowers, which have obviously been widely used in perfume and cosmetic manufacture, are an important part of the attraction exerted on pollinators. They are generally made up of a combination of fragrant molecules. These molecules are **terpenoids** (isoprene polymers) between 0.1 and 0.25 kD, essentially monoterpenes (C_{10}) and sesquiterpenes (C_{15}), and phenolic compounds and benzene derivatives (Dudareva et al., 2000a). Derivatives of fatty acids, as well as sulphurous and nitrogenous compounds, are sometimes present (e.g., amines of Araceae). The enzymes implicated in the synthesis of floral scents have not always been precisely identified; moreover, some have not yet been found in the floral tissues but only in the vegetative tissues.

Site of perfume synthesis

Perfumes are frequently produced and emitted by the same cells, located in the petals.[4] This is the case with snapdragon (*Antirrhinum majus*). However, some flowers have structures specialized in the emission of perfumes (see below). In snapdragon, both the emission and production of methyl benzoate occur in petals, as demonstrated by *in situ* hybridization experiments using the cDNA of the mRNA of S-adenosyl-methionine benzoate methyl-transferase (Dudareva et al., 2000b), an enzyme catalysing the final reaction of methyl benzoate production. In *Clarkia breweri*, the stigma also has the capacity for synthesis, as suggested by *in situ* hybridization experiments involving linalol-synthase. In both cases, it is the epidermal cells that produce the fragrant molecules. These volatile molecules escape directly. When the flower is about to fade, it no longer emits a scent. This results not from a decline in the quantities of synthesis enzymes, but from the **conjugation** of the fragrant molecule resulting in a non-volatile glycoside (Dudareva et al., 2000b).

The emission of perfume follows a **circadian rhythm** (see also below), the greatest amount being emitted during the day. It has also been supposed that the floral perfume of a particular species is specific, that is, it attracts a particular category of insects and repulses others—particularly "deceivers"—but these suppositions are yet to be proved. In any case, the floral perfumes may evolve at different times during the life of the flower: an early perfume attracts specialist pollinators, a late perfume attracts more generalist pollinators.

Specialized production structures

A remarkable and well-known example of specialized structure is that of Araceae (Vallade, 1999). In *Arum maculatum*, the floral scents are emitted by perfume-producing glands, or **osmophores**. The emission of perfume is facilitated by the production of heat, or **thermogenesis**. The inflorescence of arum is a spadix. The **spathe** surrounds an inflorescence axis that has female flowers located in the lower part and male flowers in the upper part. Below the male flowers, there are osmophores, a kind of filaments of sterile flowers. Attracted by the perfume, pollinators (dipterans) enter the spadix, reach the female flowers, and pollinate them if they have earlier visited other arums. The sterile flowers also have stiff hairs pointing downward, which more or less prevent the insects from coming out of the inflorescence until the male flowers have reached maturity (the arum is **protogynous**: the female flowers mature before the male flowers). Thus, when coming out, the dipterans are covered with pollen. The cell groups that produce perfume, the osmophores, produce heat. The male flowers produce salicylic acid, which induces a decoupling of the respiratory

[4]In Chloranthaceae, for example, where the flowers have no petals, the perfume is produced by stamens. In Solanaceae, the perfume is produced by anthers.

chain[5] and finally the production of heat. The temperature could reach about 40°C. Such thermogenesis also exists in Cyclanthaceae, Arecaceae, and Nympheaceae.

There are two types of osmophores: **epidermal** osmophores (the osmophoric cells are epidermal, as in Asclepiadaceae) and **mesophyll** osmophores (cells of the mesophyll, as in Araceae). The volatilization of perfume molecules is favoured by the protruding shape of the osmophore or even by the **papillate** (i.e., partly disconnected) nature of the osmophoric cells. Moreover, the cuticle is thinner. In so-called perfume flowers (Orchidaceae, Gesneriaceae, etc.), the osmophores occupy a large area. The perfume is recovered from that area by insect pollinators (male orchid bees), which use it as a precursor of sexual pheromones.

Ecological perspective

The relative specificity between the floral scent and the pollinator has led scientists to propose a model of speciation by evolution of perfume. In other words, a change in floral scent induced a change in pollinators, leading to the reproductive isolation of the sub-population considered and then to divergence and speciation. The specificity between the floral scent and the pollinator is for example very close in Orchidaceae, some of which synthesize pheromones that mislead the male pollinator. The pollinator may go as far as to copulate with the flower.

The capacity to produce a perfume seems to be a life trait that is easily lost during evolution because, within a set of species of a single genus, some plants produce perfume and others do not. In the genus *Clarkia*, species that do not produce perfume differ from those that do only in the linalol-synthase regulatory sequences. The difference, in that case, lies in the **regulation of the genetic transcription** and not in the possession or absence of genes coding the synthesis enzymes. Nevertheless, additional studies are still needed in other genera in order to determine whether this hypothesis can be generalized.

Some flowers, particularly Orchidaceae, "deceive", i.e., they have an attractive scent but do not offer nectar (see also Chapter 5). However, in most (other) cases, the emission of perfume seems to stop when nectar production diminishes (Dudareva et al., 2000b); otherwise the insect will probably associate the flowers with the absence of nectar and no longer visit it, ultimately to the detriment of the plant. Moreover, most often, the emission of perfume drops considerably as soon as pollination occurs, following the glycosylation of fragrant molecules (see above) and finally the drooping of floral parts.

[5]The gradient of protons created during the transit of electrons along the redox chain is not used for the production of ATP but is supposed to flow through a special protein, that produces heat and that has not been found yet. Moreover, alternative oxidase diverts electrons from the respiratory chain and prevents the formation of the protons gradient, a process that produces heat.

The composition of perfume in flowers without nectar that have a long life span or are grouped in inflorescences is also an important ecological factor. In a given species, the perfume emitted by a flower is composed of a set of fragrant molecules that is quite constant and characterizes the species, and a set of highly variable molecules that prevent the insect from learning to associate the perfume with the absence of nectar. For example, in *Ophrys sphegodes* (a species that does not have nectar), the floral perfume has about one hundred compounds, about twenty of which affect the olfactory receptors on the antenna of the pollinator. Selection pressure has an effect so that the active compounds are less variable in proportion and nature than the non-active compounds. However, the aldehydes and esters that belong to active compounds change from one flower to another, so that the pollinators will not associate the fragrant bunch with the absence of nectar. As a matter of fact, 67% of pollinators visit at least a second flower on the same inflorescence (Ayasse et al., 2000).

g) "Internal clock" and floral timing

There are two kinds of temporal control of the life processes of the flower: one related to floral **dynamism**, the other related to the **life span** of the flower. Here, dynamism refers to the processes subject to variation during the life of the flower, which includes the movements of floral organs, the emission of perfumes, and nectar production.

Dynamism and rhythm

The movements of floral organs in flowers that last for several days are rhythmic, as is the production of perfume molecules. The opening of flowers and the synthesis of perfumes are thus subject to a **circadian rhythm** controlled by an "internal clock". This term refers to a set of genetic interactions that determine in each cell a rhythmicity of genetic expressions and enzymatic activities. Even though the genes governing circadian rhythms have not all been precisely identified, some are known, such as *lhy* or *cca-1* (Fig. 4.13). The entry of the cellular oscillator (*input*) comprises the phytochrome (*phy*), sensitive to red light, and cryptochrome (*cry*), sensitive to blue light. The genetic oscillation (*lhy, cca-1*) is continually adjusted to the day-night alternation of the environment by pathways involving especially GMPc and Ca^{2+}. The genetic oscillator is at the origin of the regulation of genes downstream, that is, activations or repressions. Besides, ionic flows involving Ca^{2+} or K^+ could be responsible for the regulation of some enzymatic activities through phosphorylation events. Even though circadian regulation of the activity of perfume synthesis enzymes remains to be determined precisely, it is probably integrated within an overall process, as we will indicate further on. By analogy with the *pulvinus* of Fabaceae, it seems clear that movements of floral organs of the perianth, which are due to variations of water potential of cells, are controlled by ionic flows that depend on phytochrome and cryptochrome pathways.

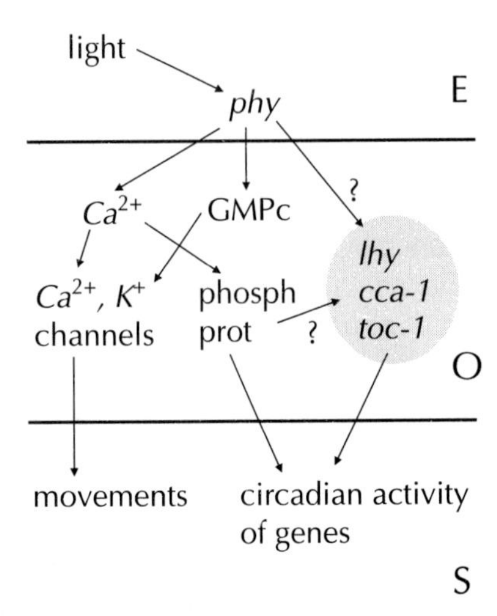

Fig. 4.13. Highly simplified diagram of determination of circadian rhythm. The cellular process of control of circadian activities is split up into input (or photoreception, E), oscillator (O), and output (S). The signals of input, light signals in particular, act on the phytochrome (*phy*) and control the release of calcium, the production of GMPc, and, by a pathway that is so far incompletely understood, the expression of genes of circadian rhythm that are *lhy, cca-1*, and *toc-1*. The quantity of mRNA of these genes oscillates during the day. This controls the expression of other genes that are dependent on them. Moreover, the calcium will lead to the phosporylation of proteins (phosph prot) by the intermediary of kinases controlled by calmodulin, which could also have an effect on the transcription of genes and the control of the activity of proteins. The ionic flows (Ca^{2+}) are the source of osmotic changes and movements called "turgescence", such as the movements of petals (or leaflets in *Oxalis*). Phytochrome, sensitive to red light, influences the activity of ionic channels in the cells of the "extensor" type, while cryptochrome, sensitive to blue light (not illustrated), induces the activity of ionic channels in the "flexor" cells. The circadian alternation of light rich in red and blue wavelengths thus leads periodically to the extension and bending of organs, such as petals, having these cell types.

A very similar process probably occurs in flowers that last only one day, since in general the hour of floral opening is not random. The constancy of floral opening and closing in a species, which has been observed since ancient times, is clearly linked to the cellular measurement of time and the existence of an "internal clock" (Table 4.2).

Table 4.2. Some examples of flower opening times

Name	Family	Opening time
Carnation	Caryophyllaceae	1.00 a.m.
Water lily	Nympheaceae	7.00 a.m.
Volubilis (bindweed)	Convolvulaceae	10.00 a.m.
Marvel of Peru (four-o-clock)	Nyctaginaceae	5.00 p.m.

The hour of flower opening and closing is probably a character subjected to selection, because the accessibility of the flower must coincide with the pollinator's period of activity. For example, in the Nympheaceae *Victoria*, it has been observed for a long time that the flower, open during the day, quickly closes in the evening, trapping coleopteran pollinators. The reopening of flowers the next morning liberates insects covered in pollen.

Fading of the flower

General points

At some time, the flower undergoes a process of drooping during which the floral architecture changes: the perianth disappears completely, in general, as do the stamens. Some organs fall and others simply die. The fall of parts of the perianth is promoted by the development of an **abscission zone**. This is characterized by the activity of degrading enzymes (cellulase, endoglucanase), which probably change over time. For example, it has been shown that several cellulases succeed one another during the abscission of a floral pedicel: early cellulases induced by ethylene and later cellulases (Campillo et al., 1996).

Physiological perspective

The drooping of sterile parts of the flower is due to **programmed cell death** (PCD). The PCD of petals has been particularly studied in morning glory (*Ipomoea* species, Convolvulaceae; reviewed in Mohr and Schopfer, 1994) and petunia. In petunia, PCD is characterized by the modification of quantities of RNA and proteins, the expression of RNases, the fragmentation of DNA, and the rupture of the plasma membrane. The double-strand and single-strand DNases are simultaneously expressed during this process and are activated by Ca^{2+}, just like the apoptotic process in animals. On the other hand, unlike the latter, the liberation of mitochondrial cytochrome *c* has not been observed in PCD of petals (Xu and Hanson, 2000). The metabolic pathways involved in the triggering and control of PCD in petals still largely remain to be studied and, in particular, it remains to be determined to what extent they can be compared to xylem PCD (Fukuda et al., 1998).

In terms of hormones, the senescence of the corolla is generally triggered by ethylene, which is produced after pollination, even before the pollen grain germinates (e.g., petunia, carnation, cyclamen, orchids). However, there are cases of floral drooping independent of pollination, as in morning glory, in which the perianth begins senescence 12 to 18 hours after the flower blooms. The involvement of ethylene has been shown simply by using inhibitors of ethylene production or even blocking its perception (the genes and enzymes implicated in ethylene production have also been determined; see box). Ethylene production is also triggered by auxin treatment, an injury to the flower, or emasculation in many Orchidaceae (*Cattleya, Cymbidium, Phalaenopsis, Vanda*).

Determination of Ethylene Production Leading to Floral Drooping

In physiological terms, the response to pollination differs according to the compatible or incompatible nature of pollination (O'Neill, 1997). In petunia, compatible pollination leads to a two-phase production of ethylene (with peaks at 3.00 and 20.00 hours), while self-incompatible pollination induces only the first phase. This has revealed that there are **two steps** (primary and secondary) in the production of ethylene, controlled at least partly in different ways. From a molecular perspective, it has been shown clearly in carnation (*Dianthus caryophyllus*, Caryophyllaceae) and petunia (Solanaceae) that genes coding ACC-synthase and ACC-oxidase* are induced after pollination, or even after an auxin treatment (Jones et al., 1999). In the same way, the mRNA of ACC-synthase are also produced during floral senescence and after the pollination of *Phalaenopsis* sp. (O'Neill et al., 1993). In this orchid, the enzymatic activity of ACC-synthase considerably increases as soon as from 12 hours and 24 hours after pollination in the stigma and in the ovary or labellum respectively. However, there are at least three genes coding for ACC-synthase (*Phal-ACS1, Phal-ACS2*, and *Phal-ACS3*), and it seems clear that *Phal-ACS1* and *Phal-ACS2* are expressed in the stigma while *Phal-ACS3* is expressed in the ovary. Besides, while *Phal-ACS1* is induced by ethylene (there is thus a positive feedback on the expression of this gene) and not by auxin, the reverse is true for *Phal-ACS2* and *Phal-ACS3* (Bui et al., 1998). Even though ACC-synthase levels are high in the ovary, ACC-oxidase and ethylene levels are very low, which suggests that the conversion of ACC into ethylene is very low in this organ. Moreover, the mRNA coding for ACC-synthase and ACC-synthase activity are very weak or absent in the perianth, a whorl that, unlike the gynoecium, enters into senescence after pollination. Finally, ethylene at high doses causes the death of the gynoecium. This suggests that there exists a translocation (the mode of which is yet to be precisely understood) of the ACC produced in the gynoecium towards the organs of the perianth.

Similar data were obtained in carnation, which contains at least three genes coding for ACC-synthase: *DcACS1, DcACS2*, and *DcACS3*. Similarly, *DcACS3*, which is expressed in the ovary, seems induced by auxin but is expressed at very low levels *in planta* (Jones et al., 1999). By contrast, the mRNA of ACC-oxidase are present in high levels in the gynoecium and it is possible that ACC is transported from other whorls to the gynoecium, which in turn forms ethylene. The ethylene induces senescence of the

*Or 1-amino-cyclopropane-carboxylic acid synthase. This enzyme, involved in the synthesis of ethylene, catalyses the reaction S-adenosyl-methionine $\rightarrow$ ACC, which is then converted into ethylene by ACC-oxidase.

perianth, while the gynoecium is stimulated towards fruit formation. Thus, in carnation as well in the orchid *Phalaenopsis*, pollination probably triggers, through auxin, the expression of ACC-synthase (of the ACS3 type) and/or ACC-oxidase in the ovary, inducing the production of ethylene (a primary step, see above). The latter affects the other whorls, exerting a positive feedback on the expression of the genes coding for ACC-synthase (the ACS1 type), thus increasing the quantity of ethylene (secondary step). This induces the senescence of the perianth and thus the drooping of the flower (Fig. 4.14). The stigma also has a similar process, the auxin stimulating at first the gene coding for ACC-synthase of the *ACS2* type. The link between pollination and auxin production remains to be understood precisely. However, in *Phalaenopsis*, the pollinia themselves contain this hormone. Finally, the presence of factors other than auxin that trigger senescence cannot be ruled out.

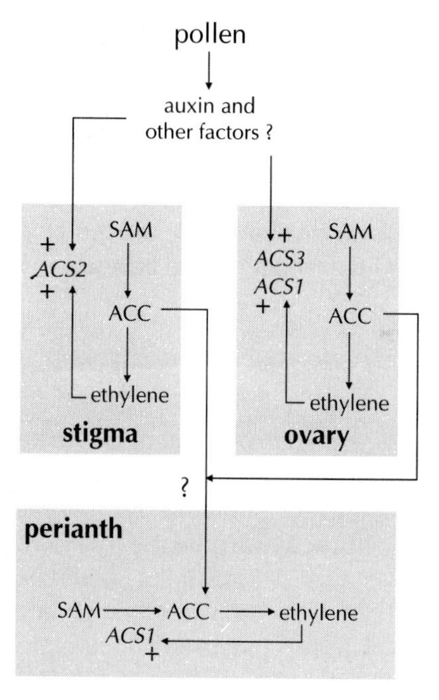

Fig. 4.14: Physiological determination of drooping in Orchidaceae. Pollination leads to an increase in the ACC-synthase (ACS) activity by increasing the transcription of *ACS1* genes in the stigma and *ACS3* genes in the ovary. The ethylene produced exerts positive feedback on the transcription of *ACS* genes (and also *ACO*, coding for ACC-oxidase, not illustrated). The ACC (and probably also the ethylene) is transported to the labellum, where it is converted into ethylene. The modalities of this transport remain to be specified (?). SAM, S-adenosyl-methionine; ACC, 1-amino-cyclopropane carboxylic acid; + activation. (Modified from Bui et al., 1998.)

In carnation flowers, abscisic acid (ABA) level increases before the beginning of the senescence process, and a pre-treatment with ABA increases ethylene evolution and sensitivity to ethylene. So ABA also plays a role in senescence of carnation flowers via its effect on the ethylene-regulating pathway.

The senescence of other flowers (Amaryllidaceae, Liliaceae, Iridaceae) does not appear to be regulated by ethylene. For example, in daylily, which is an ethylene-insensitive flower, addition of ABA promotes senescence-associated events and now some studies show that ABA is important for flower fading (for a review, see Rubinstein, 2000).

Ecological perspective

Apart from the influence of pollination, the longevity of flowers depends also on the species. Flowers of alpine species, for example, last longer than those of low-altitude species. This trait has probably been selected because of the scarcity of pollinators at high altitudes. Finally, depending on the families, there are flowers that last only one day (e.g., Convolvulaceae) and some that last several months (e.g., Orchidaceae).

In some cases the drooping of the flower leads to a cessation of nectar and perfume production, either because of a precocious glycosylation of fragrance molecules (see above), or even because it is the parts of the perianth that carry the perfume-producing cells. In any case, it is likely that the halt in the emission of perfume when the flower droops is an **adaptive** character preventing pollinators from associating the flower with the absence of reward (nectar).

h) Floral syndromes and preferences of pollinators

General points

Flowers specialized with respect to pollinators, as we have seen in the preceding sections, have some typical pollination characters. That is why, as we have suggested, some characteristic floral traits can be used as indicators of the pollinator type: this is the **floral syndrome** theory. However, it must be emphasized that in reality a whole set of floral traits should be taken into account to distinguish one mode of pollination from another, because (1) the various floral traits are not independent and (2) the pollinators are probably sensitive to several elements in a single flower. These two aspects introduce a blurriness in the notion of pollinator **preference**: what are the floral characters really used by pollinators to discriminate between flowers and what are the consequences for plants in terms of selection? Some approaches have used mutants or hybrids, for example in *Mimulus* (formerly Scrophulariaceae). Using F2 hybrids between *M. cardinalis* (pollinated by hummingbirds) and *M. lewisii* (pollinated by bees or bumblebees), it was demonstrated that an allele causing an increase in nectar or pigments of the corolla (carotenoids or anthocyanins) considerably diminished the visit rate by bees (Schemske et al., 1999; Fig. 4.15A). On the other hand, the visit rate by bees increases with the projected area of the corolla. Even though hummingbirds seem to show a preference for red (Campbell et al., 1997; see Chapter 5), this study did not show an obvious link between the flower colour and the visit rate by hummingbirds. The carotenoid concentration had a negative effect on visits by bees and no significant effect on hummingbirds: it is possible that the high carotenoid level in the corolla of *M. cardinalis* served primarily to **discourage** Apidae pollinators and not to attract birds.

As in *Mimulus*, bumblebees discriminate against small and asymmetrical flowers of *Epilobium angustifolium* (Onagraceae), which contain (significantly) less nectar (Fig. 4.15B). Fireweed plants (*Epilobium*) growing in conditions of low light or low water supply produce sparser inflorescences, with small, asymmetrical flowers containing less nectar than plants growing in better conditions. This means that the genetic variants, or more generally the individuals or clones with small, asymmetrical flowers, are not selected.

Floral syndromes and environment

The occurrence of various modes of pollination depends heavily on the **latitude**, the **altitude**, and the **forest stratum**, particularly in tropical regions. The tropical forests are structured in an undergrowth, sub-canopy, and canopy. Each stratum is marked by the presence of particular families (e.g., Rubiaceae are found mostly in the sub-canopy and undergrowth), and by particular environmental conditions, which have an effect on the modes of pollination (Table 4.3).

The modes of pollination exploiting vertebrates are relatively rare in the canopy and more frequent in the undergrowth (around 20%). Generally, it seems clear that the modes of pollination are more varied in the undergrowth than in the canopy, which probably reflects the greater species diversity of Angiosperms

---

Fig. 4.15: (A) Representation of regression coefficient obtained for the intensity of visit of flowers in relation to various parameters in *Mimulus*. (B) Representation of number of visits by insects as a function of floral symmetry in *Epilobium angustifolium*. (C) Representation of the mean curve linking the volume of nectar available to the corolla asymmetry in *E. angustifolium*.

(A) While the intensity of visits by bees is negatively related to the carotenoid or anthocyanin content and to the volume of nectar available, the insects tend to visit flowers having a larger projected surface. The opposite is true for hummingbirds. (Modified from Schemske et al., 1999.)

(B) The "shortened symmetrical" individuals have symmetrical flowers in which the lower petals have been shortened by 1 mm while the shape is preserved. The symmetrical flowers are significantly more often visited than asymmetrical flowers (modified from Moller, 1995).

(C) Individuals having a discrepancy in lower petal length that exceeds 1.5 mm produce almost no nectar. This phenomenon explains the preference of insects for symmetrical forms observed in (B). (Modified from Moller, 1995.)

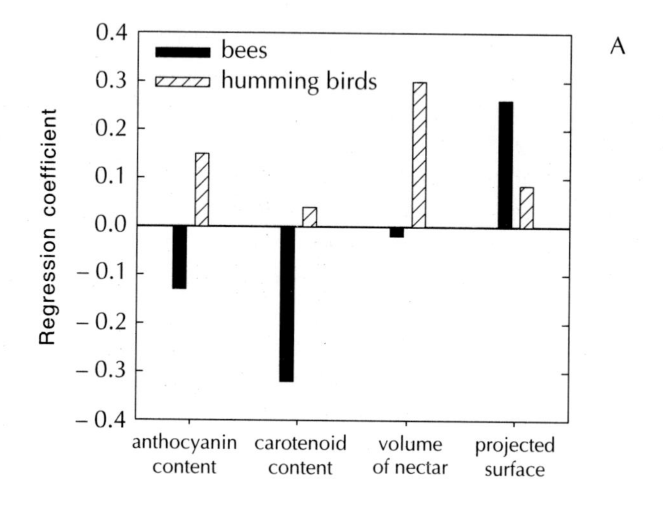
A
0.4
0.3
0.2
0.1
0.0
– 0.1
– 0.2
– 0.3
– 0.4
Regression coefficient
bees
humming birds
anthocyanin content
carotenoid content
volume of nectar
projected surface

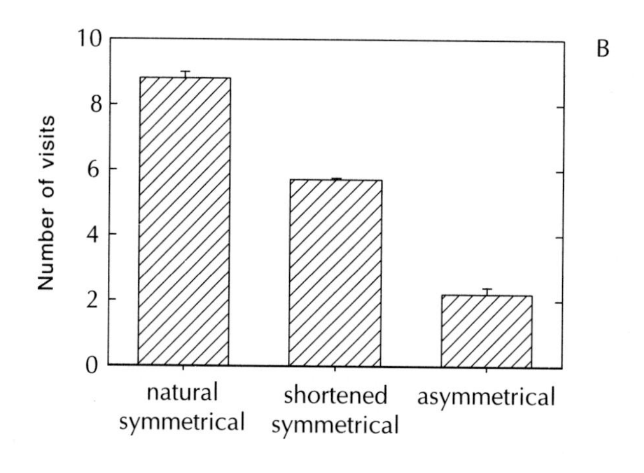
B
10
8
6
4
2
0
Number of visits
natural symmetrical
shortened symmetrical
asymmetrical

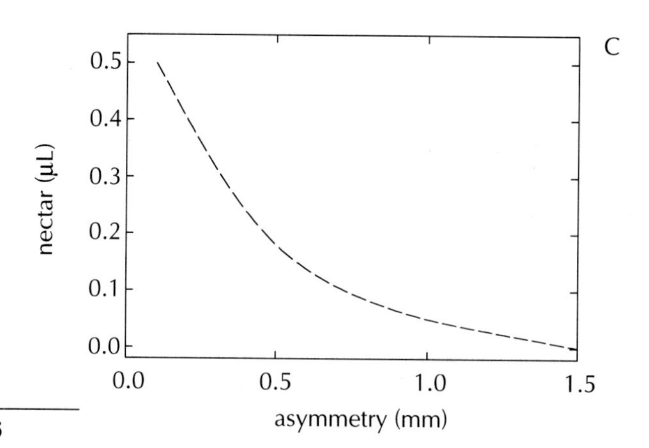
C
0.5
0.4
0.3
0.2
0.1
0.0
nectar (μL)
0.0
0.5
1.0
1.5
asymmetry (mm)

Figure 4.15

Table 4.3. Frequencies of various modes of pollination in the tropical rainforest as a function of the stratum

Pollinator	Canopy (%)	Lower canopy and undergrowth (%)
Bats	4	4
Hummingbirds	2	18
Apidate type*	52	39
Coleoptera	—	15
Lepidoptera	15	11
Wasps	4	2
Other insects	23	8
Wind	—	3

*An "apidate" pollinator belongs to Apidae, Andrenidae, Anthophoridae, Halictidae, or Megachilidae (modified from Bawa, 1990).

in the lower stratum. Remarkable, moreover, anemophily is non-existent in the canopy, which is undoubtedly linked to the continuous presence of leaves in this stratum, a factor that considerably limits pollen transport by wind.

In the temperate forests, the modes of pollination do not involve vertebrates, except in Australia and South Africa, but rather insects and wind. Indeed, the frequency of wind pollination increases progressively with the latitude (around 90% near the polar circle). Undoubtedly this can be linked with (1) the climatic rhythmicity (there are no leaves at the beginning of spring to hinder pollen transport by wind) and (2) the scarcity of pollinators at high latitudes.

The flowers of the tropical mountains are mostly ornithophiles (hummingbirds) and generalist entomophiles, since high altitude, like high latitude, is accompanied by a scarcity of insects. Similarly, chiropterophilous pollination is rarer at high altitude. Finally, probably because of the unpredictability of pollination, mountain flowers (particularly in the tropics) bloom longer and self-compatible tree species are more frequent.

4.1.3. Dispersal and Collection of Pollen

a) General points

Whatever the mode of pollination is (biotic or abiotic), the pollen dispersal and collection (by the stigma) are **linked**, because they are related to the agent of pollen transport. With respect to the biotic modes of pollination, there are at least two types, from the pollinator's point of view: one **passive**, the other **active**. In the passive mode, the pollinator deposits or collects pollen on its own body without its "knowledge". In this case, the fertile parts may or may not be animated by a movement. In the active mode, the insect has to exert a particular movement that leads to the deposit or collection of pollen. The distinction between these two modes is sometimes fluid because it is difficult to know from what moment we can say that the behaviour of the pollinator is modified. For example, the floral attraction already modifies the pollinator's behaviour to the extent that it affects the pollinator's trajectory. We confine ourselves to distinct cases of active pollination in which the pollinator makes a movement other than pure and simple search for nectar. Finally, we will deal with one major type of biotic pollination, **mutualism**, in which the reproduction of the pollinator is linked to that of the plant.

In general, wind-pollinated flowers have well-exposed stamens, with elongated filaments (see above). The stigma has a large area of reception, developing a feathery shape (e.g., Poaceae). The flowers sometimes have a very small perianth, which reduces the hindrance to wind transport of pollen grains (e.g., Salicaceae). The floral characters linked to hydrophily, which are quite similar, have been discussed earlier (see section 4.1.1).

b) Passive biotic pollination

The various kinds of passive biotic pollination can be classified not only by the presence or absence of movement of floral parts, but also by the nature of the floral parts involved in such a movement (sterile or fertile).

A typical example of passive pollination that does not involve floral movement during pollination is that of many Orchidaceae (see above). There are numerous other examples, including among the ornithophilous flowers. Many flowers that are not specialized in terms of pollination are classified among the plants with passive pollination (e.g., Ranunculaceae, Magnoliaceae).

Plants with passive pollination and exerting a floral movement include taxa belonging to Fabales as well as Lamiales and generally have zygomorphic flowers. Sage (*Salvia* sp., Lamiaceae) is a typical example of an arrangement involving "pedal" stamens (see above). Fabaceae, as we will now see, are also a remarkable case of mechanism suitable for explosive movement.

The order Fabales comprises two families abundantly represented in Western Europe: Fabaceae and Polygalaceae. Flowers of Fabaceae and Polygalaceae are zygomorphic and of type 5. In Fabaceae, the abaxial part of the corolla, composed of two united petals, is called the **keel**. The upper petal, the **banner** (or standard), clearly visible and having a central groove, is not the result of a union of two petals: it is a single large petal with an arrangement that is the reverse of that observed otherwise in zygomorphic flowers, i.e., adaxial. The two lateral petals, the **wings**, enclose the keel. Finally, the stamens may be joined by their filaments, the androecium being thus structured in 9 united abaxial stamens + 1 adaxial stamen (**diadelphous** androecium). A flower with such architecture is called "**papillionaceous**" (Fig. 4.16). The **inverted**

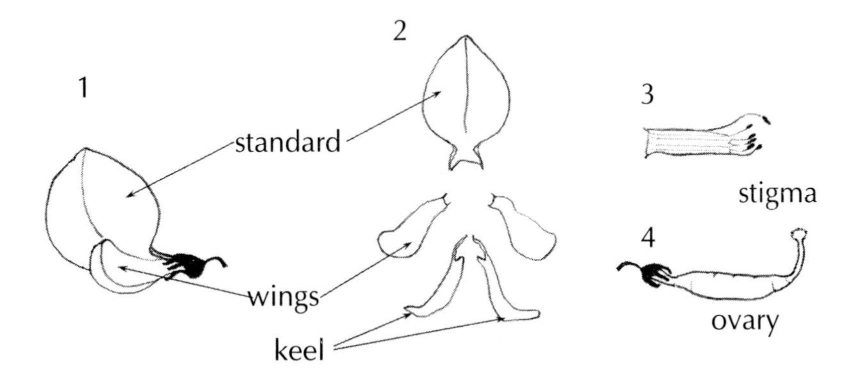

Fig. 4.16: Diagram of a papillionaceous flower of Fabaceae. (1) Outer view. (2) Dissected corolla. (3) Isolated androecium seen in profile, showing its didynamous organization into 9 + 1 stamens. (4) Flower in which the corolla and androecium have been removed in order to reveal the gynoecium.

Mode of Pollination in Other Fabaceae

The description given above does not apply to all Fabaceae: although this architecture is observed in Faboidae, Mimosoidae present actinomorphic flowers with a reduced corolla and more than 10 stamens with very long filaments. Besides, the flower is not inverted. In Caesalpinioidae,* the floral structure is similar to that of Faboidae, except that the adaxial petal is internal and not external, and the two abaxial petals are not united.

Caesalpinioidae have remarkable examples of pollination by Lepidoptera. In *Caesalpinia*, the filaments of the stamens and the style are very long, vivid, and slightly erect (Fig. 4.17). The corolla is intensely coloured. The pollen is liberated from the anther in the form of sticky aggregates. The stigma comprises a depression surrounded by trichomes. The nectar is produced by a nectary at the base of the gynoecium, and it is accessible only through the tube formed by the adaxial petal. The trichomes of filaments prevent access to the centre of the flower. The insect approaches the flower and engages its proboscis in the tube of the upper petal, while continuing to beat its wings because there is no "landing" area. While the insect hovers, the pollen from the anthers sticks to its wings and the pollen carried by the insect is received by the stigma.

Mimosoidae such as *Acacia* sometimes have a large number of stamens. The flowers are arranged in dense, often globular inflorescences (Fig. 4.17). The flowering leads to very high production of pollen, potentially leading to geitonogamous self-fertilization. In fact, these species are observed to be protogynous and also self-incompatible (see gametophytic self-incompatibility below). The pollinators are Hymenoptera looking for pollen. At least some *Acacia* of Australia have **extrafloral nectaries** that attract birds.

*This subfamily is a paraphyletic group.

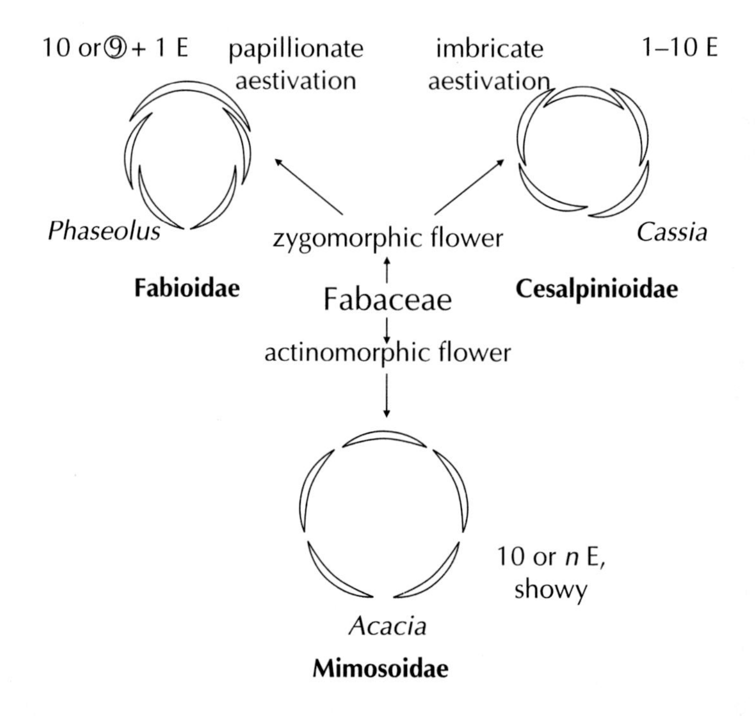

Fig. 4.17: Key to Fabaceae using floral criteria. For each sub-family, a genus is cited as an example. The three diagrams illustrate the form of the corolla, revealing its aestivation. This diagram does not represent a phylogeny; moreover, the group "Cesalpinioidae" is paraphyletic.

arrangement (see Chapter 3) of the flower puts the keel in the lower position. The keel, folded like a boat, leads the anthers and stigma to touch the body of the pollinator: deposit and collection of pollen are thus coupled (e.g., in *Genista, Medicago*).

c) Active biotic pollination

Flowers having a structure that constrains the pollinator to make a particular movement are found in Monocotyledons (e.g., Iridaceae, Araceae) as well as in Dicotyledons. These flowers sometimes produce nectar and have a particular architectural arrangement that obliges the insect to change its posture. Contrary to what might be expected, many Orchidaceae are classified among flowers with passive pollination: the pollinia are deposited on the body of the insect without its "knowledge", while it searches for nectar.[6]

[6]In some cases, however, the insect is forced to follow a particular path in the flower, as in *Cypripedium*, which is included among flowers with active pollination. See also vanilla (above).

For example, in iris, there are **ornamentations** on the tepals that, apart from their probable role of attraction, are erect like hairs, obliging the insect to lift its abdomen as it progresses into the flower. That brings it in contact with the stamen. A similar strategy exists in Lamiaceae such as ground ivy, in which the lower labium has erect hairs (Fig. 4.18).

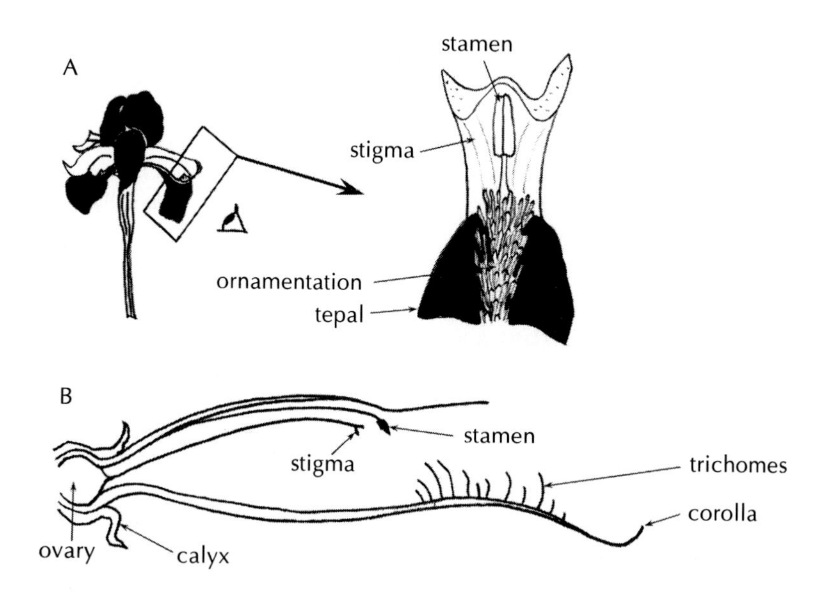

Fig. 4.18: Diagram of an iris flower (A) and a section of the flower of ground ivy (B). (A) The general appearance of the flower is shown at left and the detail of the interior of the flower seen at a low angle (indicated by the eye) at right. Note the large dilation of the sigma.

d) Mutualism

General points

The plant-pollinator interactions linked to reproduction imply an interdependence of the two species, that is, without one of the partners, the reproduction of the other is impossible. This is typically the case with **fig** (Moraceae), **globeflower** (*Trollius*, Ranunculaceae), and **yucca** (Agavaceae), which have been abundantly studied in the past few years from a naturalist and theoretical perspective. The case of fig has been treated in other studies (Vallade, 1999) and the reader is referred to those studies. However, note that, unlike yucca and globeflower, fig pollen is dispersed by wasps emerging from the fig and thus the wasps offer something in return for their growth in the gynoecium (Anstett et al., 1997). In other words, while the fig is selected to "grow up" the wasp progeny, globeflower and yucca are *a priori* selected to kill the progeny. Here we will discuss only the example of yucca.

Yucca

Yucca is pollinated in association with Prodoxidian Lepidoptera or "yucca moths". These insects use specialized mouth parts to collect pollen and deposit it on the stigma of the next flower, after having injected eggs into the ovary. The larva, still enclosed in the pistil, feeds on the seeds during its development. Once sufficiently grown, the larva leaves the ovary, falls on the ground, buries itself in order to hibernate, and eventually enters into diapause. When conditions are again favourable, the larva enters metamorphosis; the moth leaves the ground and mates in a *Yucca* flower. Once it has mated, the female collects the pollen and flies to another flower. The adult (imago) generally does not feed and dies after three or four days. Three genera of insects are associated with *Yucca* (Table 4.4): *Prodoxus, Tegeticula*, and *Parategeticula*. It is not a one-to-one relationship (one species of *Yucca* for one species of pollinator), since one species of *Yucca* has a set of pollinator species with a more or less marked mutualism.

Table 4.4. Genera of insects that have a more or less mutualist relationship with *Yucca* (modified from Powell, 1992)

Genus	Example of *Yucca* partners	Pollen transport	Position of eggs
Prodoxus (~ 10 sp.)	*Y. arizonica, Y. baccata*, etc. (~ 10 sp.)	No	In sterile tissues of *Yucca*
Tegeticula (3 sp.)	*Y. arizonica, Y. baccata*, etc. (~ 10 sp.)	Yes	In the carpellary locules
Parategeticula (1 sp.)	*Y. schottii, Y. elephantipes*	Yes	In the stalk or perianth

The **floral architecture** is slightly modified in relation to other Agavaceae: the ovary is enlarged, the stamens are short (Fig. 4.19). The flowers are protogynous and self-incompatible. These characters considerably limit the possibilities of self-fertilization. The flowers produce very little nectar, which limits the attraction exerted on other non-specific pollinators. Finally, this association is made possible by synchronization between the emergence of imagos and the blooming of the flowers.

The **investment of the plant** in terms of number of flowers seems to be maximal, since the number of pollinators is apparently not limiting. The prevalence (rate of infestation of ovaries by larvae) is about 0.5 to 6 larvae per *Yucca* fruit. The proportion of lost ovules is less than 30% but sometimes reaches 60% (*Y. filamentosa*). Thus, the cost of the association is **highly variable** in terms of seeds sacrificed. Moreover, from one species to another, the behaviour of the insect is not always the same. Some species inject an egg in all the

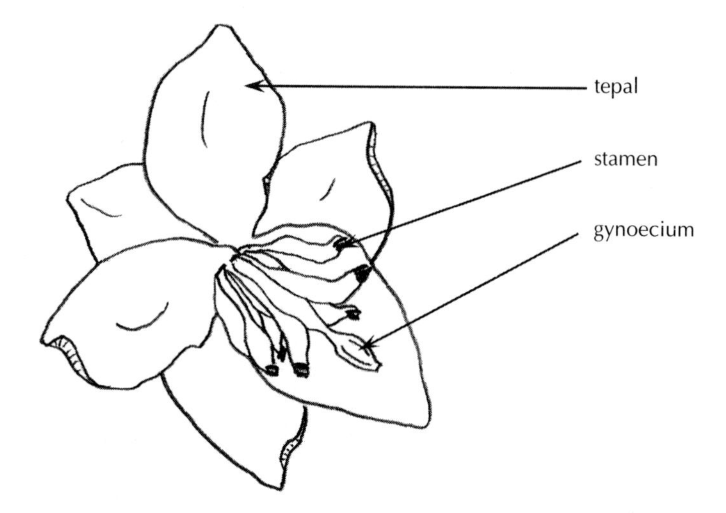

Fig. 4.19: Diagram of a *Yucca* flower in which the perianth has been slightly flattened in order to reveal the fertile parts. Note the squat appearance of the stamens.

carpellary locules (that is, 6 eggs per flower), and others limit themselves to one or two eggs per flower: the loss of seeds in a flower is thus also highly variable (Powell et al., 1992).

There are some local varieties of a given species of *Tegeticula* that do not have the requisite mouth parts to collect pollen (Pellmyr et al., 1999). These animals nevertheless exploit the plant and inject an egg in the pistil: these are called **cheating moths**. Similarly, there may be varieties of *Yucca* that do not permit the development of the larva inside the pistil; such *Yucca* varieties benefit from the pollination without losing seeds (cheating yuccas). The development of cheats threatens the stability of the association, because the cheating variants offer nothing in return and do not suffer any cost in the association, which means they will temporarily have better reproductive success than the non-cheats and outnumber the latter. How can we therefore explain the stability we observe in the association? The mathematical development of a system with two mutualist partners in the form of differential equations seems to indicate that there are only two possibilities, as a function of resources offered by the partners: coexistence or the disappearance of the system (one mutualist does not invest enough, the system is doomed to extinction). Subsequently, depending on the symmetrical or asymmetrical nature of the **coefficients of competition** for access to resources offered reciprocally by the partners, the system evolves towards extinction (symmetrical) or converges towards equilibrium (asymmetrical) (Ferdy et al., 2002; Law et al., 2001).

4.1.4. Pollination and Mate

The floral strategies of allogamous plants by definition favour allogamy, i.e., the fertilization of the stigma by a pollen grain coming from another plant. Apart from the strategies developed to attract pollinators, there are strategies aimed toward a more direct control of the pollen origin, and particularly toward the avoidance of strict or geitonogamous self-fertilization. These strategies finally lead to a kind of **choice of mate**.

a) Anatomical effects

The arrangement of floral organs may especially limit the possibilities of self-fertilization, either by the existence of an obstacle between stamens and stigma (**herkogamy**) or because the organs do not mature at the same time (**dichogamy**). The flowers are at first male (**protandry**) or female (**protogyny**). A typical case of herkogamy is found in iris, in which the style-stigma arrangement, very large and trifid, separates the stamens from the receptive area of the stigma, or in Orchidaceae (see above). **Tiered** architecture, as in the passion flower, separates the anthers (facing downwards) from the stigma (see Chapter 3).

Dioecy (existence of separate male and female plants) and thus the sexualization of the organism can be related to strategies of avoidance of self-fertilization and this point will be discussed further on. Without going to this extreme, some species have two **floral morphs**, one directed towards the male function, the other towards the female function, a phenomenon called **heteromorphy**. Primrose (Primulaceae) is a well-known example. The flowers directed towards the male function have long stamens extending beyond the stigma in the floral tube (Fig. 4.20). The stigma has no papillate surface. The

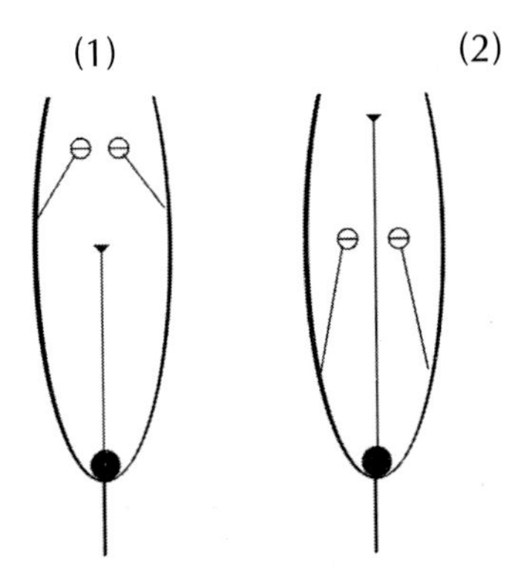

Fig. 4.20: Highly simplified diagram of floral morphs in primrose. (1) male morph, (2) female morph. The male morph has long stamens and a short style, and the reverse is true in the female morph (see text).

pollen grains are large and richly ornamented. The morphs directed towards the female function have a stigma extending much higher than the stamens, and a small and slightly ornamented pollen grain. The system tends to favour the transfer of pollen from the male morphs to the female morphs. The existence of two types of flowers, one with a long style and the other with a short style, as in *Primula*, is called **heterostyly**. There are intermediate cases in which only the difference in style length is observed, and it is not accompanied by stamen or pollen changes, as for example in some varieties of daffodil (*Narcissus* sp., Amaryllidaceae), where there is simply **distyly** (Arroyo et al., 2000). The heterostyly of primrose is accompanied by a mechanism of molecular control, the basis of a system of **self-incompatibility**, which is discussed in the next paragraph. Finally, it should be noted that although self-incompatibility is reinforced by heteromorphy in primrose, the same does not apply in some **homomorphic** plants such as Brassicaceae (see also Table 4.5).

Table 4.5. Self-incompatibility systems in Angiosperms and their determination in terms of formal genetics (modified from Heslop-Harrison, 1975)

Morphology	Type	Genetics	Example
Homomorphic	SSI	1 (or more) loci with *n* alleles; various degrees of dominance	Brassicaceae
Homomorphic	GSI	1 (or more) loci with *n* alleles; co-dominance	Solanaceae
Heteromorphic	SSI	1 locus with 2 alleles, one being dominant	Primulaceae

b) Cellular and molecular effects: self-incompatibilities

Instead of describing in detail all the molecular processes involved in self-incompatibility, which do not directly concern floral architecture strictly speaking, we will limit ourselves to presenting the essential available data about this subject. Self-incompatibility means the **rejection of "self"**, self being represented by proteins coded by genes of **locus S** (for self-incompatibility). When a pollen grain has a determinant similar to the stigma, there is a reaction of rejection, or **self-incompatibility response**.

First of all, self-incompatibility of the **sporophytic** type (SSI) or cabbage type (e.g., Brassicaceae, Asteraceae) must be distinguished from that of the **gametophytic** type (GSI) or tobacco type (e.g., Onagraceae, Liliaceae, Solanaceae, Fabaceae). In descriptive terms, in SSI, a pollen grain incompatible with a stigma does not germinate or germination stops very quickly. Moreover, in the Brassicaceae, the pollen tube of an incompatible grain does not always pierce the stigmatal cuticle. Generally, SSI is characterized by the very rapid **appearance of callose** once germination begins. In the case of GSI, the pollen grain clearly begins to germinate and then stops, while the pollen tube is in the style (except in Onagraceae, where the pollen grain does not germinate). The blocked tubes have a parietal thickening (e.g., petunia) or even rupture at their tips (e.g., tomato) (Heslop-Harrison, 1975). In genetic terms, the determination of self-incompatibility lies in the locus S genes: the two alleles of the mother plant for SSI, and the single allele carried by the pollen grain for GSI. In molecular terms, the mechanisms of sporophytic and gametophytic incompatible reactions are not identical and are discussed below (see box).

It has been proposed that there is a correspondence between GSI and wet stigmas and between SSI and dry stigmas (Dickinson, 1975; see also Chapter 2), as (1) plants with SSI have a dry stigma and (2) the behaviour of the pollen tube from an incompatible pollen grain is related to the nature of the stigma: germination begins in the abundant secretion of wet stigmas, and ceases early without piercing the cuticle in dry stigmas. However, the correspondence is not exact, since for example Graminaceae (Poaceae), plants with GSI, also have a dry stigma.

Finally, note that although the homomorphic plants have SSI as well as GSI, the self-incompatibility system of known heteromorphic plants is exclusively sporophytic (Table 4.5). This is generally determined by a locus with only two alleles, one of which is dominant.

c) Floral sexualization

Definitions

Most Angiosperms (90%) have **hermaphrodite** flowers, i.e., flowers simultaneously having male and female parts, but some plants (10%) have unisexual flowers. **Dioecious** species have male plants with staminate flowers and female plants with pistillate flowers (e.g., mercury, Euphorbiaceae). The **monoecious** species are made up of individuals simultaneously having male flowers (staminate) and female flowers (carpellate). There are intermediate cases: **andromonoecious** species simultaneously have male and hermaphrodite flowers, **gynomonoecious** species have female and hermaphrodite flowers.

Determination

Floral architecture is especially affected by floral sexualization since some floral parts are absent. However, the floral development of unisexual flowers shows that the primordia of the two types of fertile parts are present in the early stages and that, subsequently, one type aborts (or forms sterile parts, i.e., staminodes or pistillodes), while the other develops normally. That shows that the plan of floral organization is originally hermaphrodite (or "**bipotential**"), and that the determination of that plan is accompanied by the programmed death of stamens or carpels (sexual reversions are also possible).

In terms of genetics, there is a wide diversity and sometimes a great complexity of mechanisms controlling the flower sex. The white silene, a dioecious plant (*Silene latifolia* ssp. *alba*, Caryophyllaceae) has sex chromosomes of the

Molecular Determination of Self-incompatibility

The molecular processes that allow the plant to discriminate self from non-self have been closely studied in the genus *Brassica*, the only group in which the pollen and stigma determinants of SSI are simultaneously known. The self-incompatibility response is manifested by the inhibition of hydration and the germination of the pollen grain. In genetic terms, there are around 50 known alleles at locus S in this genus. Recall that the phenotype of the pollen grain is determined by the two haplotypes S of the mother plant inherited from the tapetum cells (the self-incompatibility is sporophytic). The ability of the stigma to recognize the self-pollen grains develops around one day before the flower blooms.

Various agents of SSI have been proved, but there are uncertainties, at least in some cases, about their roles, which is why our discussion is based on experimental data. The first molecule that has been identified in the stigmatal cells is SLG (*S locus glycoprotein*), present in the cell walls of stigmatal papillae. At the plasma membrane, these cells carry the trans-membrane protein SRK (*S locus receptor kinase*). The extracellular domain of SRK has a sequence that is quite significantly similar to that of SLG. The definitive implication of these proteins in SSI has been shown in self-fertile plants, which exist spontaneously in nature, or even more recently with transgenic plants expressing another SLG/SRK allele. For example, the expression of transgene arising from the cDNA of SRK_{28} in stigmas $S_{60}S_{60}$ has led to the rejection of pollen grains produced by $S_{28}S_{28}$ homozygotes. On the other hand, while the self-pollination of $S_{28}S_{28}$ individuals led to no seed, the transgenics produced a very small number of seeds, indicating a weak self-incompatibility response. The "strong" self-incompatibility response can be observed integrally in the double transgenic $SRK_{28}SLG_{28}$ (Takasaki et al., 2000). Such experiments suggest that SRK probably represents the protein determining the specificity in the stigma, and that SLG only contributes to the self-incompatibility response.

Several roles have been proposed for SLG. First of all, since pre-treatment of stigmas with an anti-SLG antibody reduces the force of adhesion of the pollen grain on the stigmatal surface, it has been supposed that SLG was implicated in the binding of the pollen grain on the stigma. However, mutants with a deletion of the SLG gene have a normal pollen deposit and germination. Mutants with a deletion of SLG and homozygous for an allele S do not form the SRK protein, even though they produce the corresponding mRNA. Moreover, the overexpression of the SRK gene in a plant having other SRK haplotypes leads to the formation of aggregates of the protein derived from the expression of the transgene (Nasrallah, 2000). This disappears when the associated SLG transgene is also introduced. These observations agree with the assumption that SLG is involved in the post-translation processes of SRK production.

The SRK protein, a stigmatal determinant, is a receptor-kinase. The extracellular part, similar to SLG, which is the receptor part, is the variable part (as a function of alleles). The intracellular part bears the kinase activity. However, the substrates of SRK are not known. While the gene SLG contains only one exon and the protein SLG is present in only one form, the mRNA SRK can undergo an alternative splicing that leads to two forms, SRK or eSRK, the latter comprising only the extracellular domain of SRK. The role of eSRK is still to be discovered. Finally, the pollen determinant is a protein, SCR (*S locus cysteine rich*), expressed exclusively in the anthers during the development of the pollen grain (Nasrallah, 2000). The protein SCR, rich in cysteine, is secreted. As described earlier, the use of transgenic plants has shown that SCR is the factor that bears the necessary and sufficient pollen identity. The SCR-SRK linkage, perhaps stabilized by SLG, probably causes the phosphorylation of unknown substrates, which induces the self-incompatibility response (Fig. 4.21). The SRK molecules linking SCR seem to be present in dimeric form (Giranton et al., 2000).

A gene likely to be involved in the downstream processes of the SLG-SRK is the *mod* gene, coding for a protein having a great similarity of sequence with aquaporins or "water channels" (Ikeda et al., 1997). A mutation in this gene leads to the disappearance of self-incompatibility. Besides, it has been shown using reporter genes that the promoter of *mod* is active in the cells of stigmatal papillae (as well as in other vegetative tissues). The expression of *mod* in the ovocytes of *Xenopus* shows that the Mod protein increases the hydric permeability of the plasma membrane, which means that Mod is probably an aquaporin functional in the plant (Dixit et al., 2001). The phosphorylation cascade initiated by SRK leads probably to phosphorylation of Mod and, by analogy with other known aquaporins (Kjellbom et al., 1999), it lowers the conductance of water and prevents the hydration of the pollen grain.

Finally, the molecular data about the GSI are less abundant and the processes are not well known. This type of self-incompatibility exists in Papaveraceae and Solanaceae, which include two study models: poppy (*Papaver rhoeas*) and tobacco (*Nicotiana tabacum*) respectively. In tobacco, the S locus gene codes for an RNase, secreted by the cells of transmitting tissue. The RNase is taken up by the pollen tube (a phenomenon that is still unclear) and there destroys the rRNA if it is from self-pollen. In poppy, on the other hand, instead of RNase activity, DNase activity has been detected (Jordan et al., 2000), but the data are still incomplete. The DNase activity is triggered by an influx of Ca^{2+} into the pollen tube. Another consequence of the influx of calcium is the hyper-phosphorylation of the p26 protein, a pyrophosphatase, which reduces its catalytic activity. The substrates of p26 are otherwise unknown. In Papaveraceae, the product of S locus is a secreted glycoprotein, the implication of which in GSI is poorly understood.

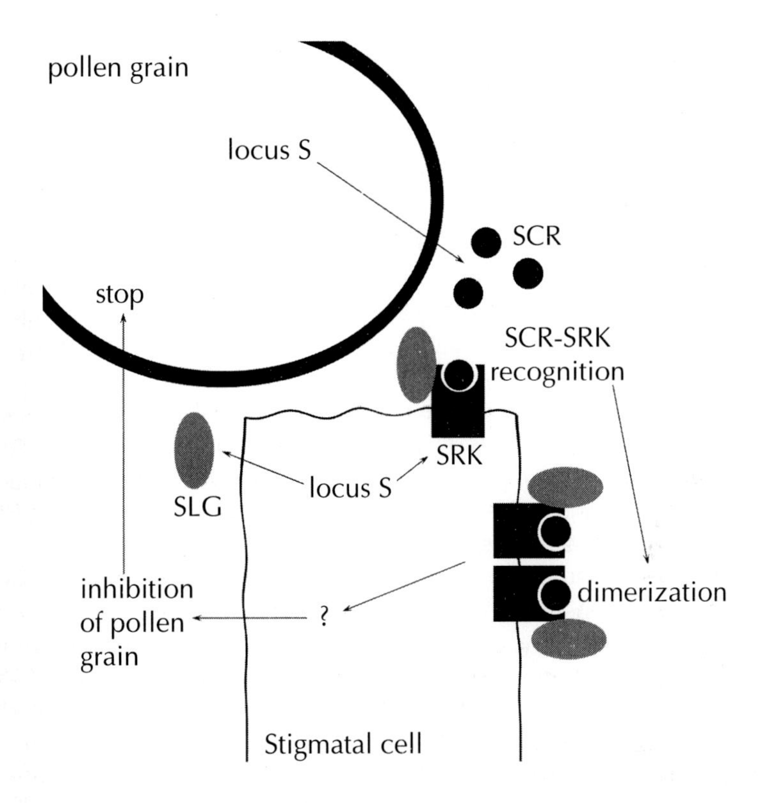

Fig. 4.21: Molecular mechanisms of incompatibility of *Brassica* type. The pollen coat has the SCR protein, coded at the locus S. On its side, the stigmatal cells have a parietal protein, SLG, and a transmembrane protein, SRK, which is a receptor of SCR. The recognition of SCR by SRK, undoubtedly aided by SLG, is followed by the dimerization of the complex, the transphosphorylation of SRK, and probably the phosphorylation of unknown cytoplasmic proteins. By a transduction pathway that is still to be specified, this leads to the inhibition of germination of the pollen grain. (Modified from Nasrallah, 2000.)

Drosophila type (XX females and XY males). It has been shown in this species that two important genes implicated in the floral sexualization are *SLM2* and *3*. These two genes are expressed in whorls 2 and 3 (petals and stamens). This expression is central in the male, which prevents the development of carpels that normally occupy this place. The expressions of *SLM2* and *SLM3* are more peripheral in the female, which results in the formation of the gynoecium. In maize, a monoecious plant, the gene *tasselseed 2* stimulates the programmed cell death of carpel cells, and its mutation is feminizing. Finally, there are species in which the individuals are either dioecious or monoecious, as in Cucurbitaceae. In *Ecballium*, for example, the dioecious-female/dioecious-male/monoecious character is simply monofactorial with 3 alleles.

However, the sex is also determined by hormones, such as **gibberellins**. For example, in maize, the mutation *anther ear 1*, affecting the synthesis of gibberellins, is masculinizing. In cucumber (*Cucumis sativus*), a monoecious plant, the spraying of gibberellins is also masculinizing. This plant is a useful study model in sex determination, because the expression of sex occurs in three phases: the early flowers are male, the intermediate flowers are hermaphrodite, and finally the late flowers are female. In terms of genetics, the sex is controlled by 3 loci with 2 alleles: *M, F*, and *A*. *F* controls feminization. The wild plants are *ff MM*. The *FF MM* or *FF Mm* individuals have only female flowers, while the *ff mm* plants are andromonoecious. The allele *F* is thus feminizing and epistatic on *M*, while the allele *M* is masculinizing. The recessive allele *a* accentuates the male character and is epistatic on *F*: the *aa ff* organisms are dioecious males (bearing only male flowers).

Apart from the influence of size on floral sex (see Chapter 5), hormones seem to be very important factors *in vivo*. The gibberellins have a masculinizing action and **auxin** a feminizing action, like **ethylene**. The involvement of ethylene in cucumber is in this respect remarkable: the dioecious female individuals produce more ethylene at the inflorescence level than the monoecious ones, which in turn produce more ethylene than dioecious males (Rudish, 1976). It has been observed that ethylene exerts a positive feedback on its production (more precisely, on the transcription of the gene coding for ACC-oxidase) at the level of apex females and triggers the orientation of the flower towards the female sex. The apex males, on the other hand, are subjected to a negative feedback of ethylene on its own production, which extinguishes its production, leading to the abortion of female organs (Kahana et al., 1999). The pathway of transduction through which ethylene exerts negative or positive feedback on male or female organs is unknown. The gene coding for ACC-synthase (ACS) is located at locus *F* and it seems that the gene coding for ACC-oxidase (ACO) is at least linked to *F*. That corresponds closely to formal genetic data, since a direct relationship is established between *F* and ethylene, a feminizing hormone. Nevertheless, the correspondence between the alleles *F* or *f* and the different *ACO* alleles is not very clear.

4.2. AUTOGAMOUS PLANTS

Autogamous, or self-fertilized, plants are rarer than allogamous plants and the strategies directed toward this mode of pollination are less numerous than those directed toward allogamy.

4.2.1. Cleistogamy

Cleistogamous flowers do not open at maturity (e.g., Violaceae). In this situation, the stamens remain close to the stigma and a pollen grain can germinate directly in contact with the stigma. However, sometimes the pollen grains are not even liberated from the anther locules, particularly in the cleistogamous flowers of Malpighiales. The pollen grains begin to germinate while they are in the anther. The pollen tube pierces the wall of the anther (Violaceae) or even grows into the filament (Malpighiaceae) and sometimes reaches the receptacle before ultimately reaching the ovule (Callitrichaceae) (see also Chapter 2). In the violet, only the late flowers are cleistogamous, while there are strictly cleistogamous species as in Onagraceae. In Poaceae (e.g., wheat), the flowers open only after they are self-pollinated.

4.2.2. Other Cases

Other plants are autogamous when they are open, for example agrimony (*Agrimonia* sp., Rosaceae), the stamens of which have a **filament curved back** towards the inside, so that the pollen falls directly on the pistil when the flower is shaken. Other flowers have stamens grouped in a bundle through which the pollen-covered style passes during its growth (e.g., Solanaceae, some Asteraceae). In Malvaceae, there are examples of **facultative** autogamy. Stigmas are above the stamen tube and have the capacity to curve downwards, causing the stigma and anthers to come into contact, if cross-pollination has not occurred (Fig. 4.22). This curvature is reversible in some species, to the extent that the stigma becomes erect if cross-pollination occurs during the curvature. The cleistogamous flowers of Malvaceae also have a "curved-back" formation.

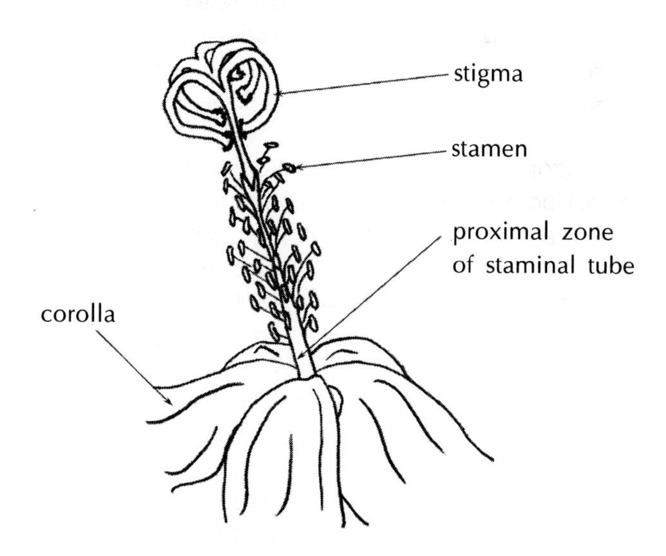

Fig. 4.22: Diagram of a facultative autogamous flower of Malvaceae.

5

Towards a Global Theory?

Angiosperms are extremely diverse, and this variety results from a radiation of this group since the Tertiary Era. In the preceding chapters we have looked at certain strategies adopted by Angiosperm species in order to realize sexual reproduction, such as strategies to attract pollinators. In the case of biotic dispersal of pollen, vectors (e.g., insects, birds, bats) intervene in the processes of natural selection that lead to fixation of adaptations at the flower level. In **qualitative** terms, it can be supposed that pollinators exert selection pressure on plants that benefits pollination. On the other hand, it is inherently difficult to **quantify** the effects of this selection on plants and, for that purpose, the phenomena at work need to be dissected into discrete components. With respect to biotic pollination, for example, selection works at several levels: e.g., attraction exerted by the flower, accessibility of stamens, deposit of pollen on the vector, possible reward (nectar) given by the flower, possible loss of pollen by the vector.

In our discussion of the selective processes that lead to the evolution of the flower, we will focus on zoophilous pollination, using certain mathematical models that allow a quantitative approach through calculations of strategies that optimize fitness (see below). The evolutionary processes can be organized in a “loop” of actions and **feedback** (Metz et al., 1992). The individuals that present particular traits determine the dynamics of the population as a whole, which in turn determines the selection pressure(s). These pressures then determine the traits of individuals and the loop is thus closed. This last transition, that is, the selection pressures and certain models that represent their effect, are the focus of our discussion. Individuals having particular life traits have a certain reproductive success, which we refer to as *fitness*.

We will thus give an overview of the theoretical framework of probable mechanisms of modification of floral characters by dissecting the various components influencing reproduction, i.e., dispersal and its consequences, aspects of sexuality and fertility, and the possibility of hybridization.

5.1. DISPERSAL AND DIVERSIFICATION

The origin and diversification of Angiosperms have puzzled biologists at least since Darwin's time. Their proliferation on the global scale has frequently been related to characteristics of their sexual reproduction, such as pollination and the dispersal of fruits or seeds. To what extent can these parameters be at the origin of Angiosperm diversification?

It has been assumed that the biotic dispersal of pollen (and seeds) increases the probability of speciation and decreases the probability of extinction. Indeed, biotic pollination could be involved through some reproductive peculiarities arising from geographic variations or behavioural variations of pollinators. The plant-pollinator co-evolution may lead to reciprocal changes that, in combination with processes of sexual selection and genetic drift, accelerate diversification. In addition, pollination by animals could favour cross-fertilization even when the plant density is low, thus reducing the probability of extinction (Real, 1977). The seeds dispersed by a biotic agent most often reach sites that have fewer pathogenic agents or consumers (Tiffney and Mazer, 1995). Finally, biotic dissemination could promote diversification by divergence (geographic speciation), when the seeds are transported sufficiently far away.

Does biotic dispersal effectively favour diversification? Some studies have led to contradictory results: some conclusions drawn from the study of Saxifragales or from 147 families of Angiosperms corroborate this hypothesis, but other studies report that families with biotic dispersal of seeds do not seem to have more species than those with abiotic dispersal (for a review, see Bawa, 1995).

However, comparisons require a certain equivalence between species, because differences (in the mode of growth, for example) could mask contrasts of diversity between groups. Studies limited to ligneous Dicotyledons seem to show a positive effect of biotic dispersal on the species diversity, according to Tiffney and Mazer (1995). These authors argue that this effect is related to the fact that trees have relatively large seeds that are able to develop in the undergrowth, where the light is low, using their reserves. Large seeds cannot be transported efficiently by wind, which explains why trees have biotic dispersal, often by means of vertebrates. Dispersal by vertebrates, generally over long distances, also prevents competition with the mother plant. It is hence probable that woody species present greater species diversity because of a lower probability of species extinction.

Some studies tend to show that species richness is correlated positively with the extent of the distribution area and that it is greater in groups with biotic *and* abiotic dispersal of seeds than in groups having only a single mode of dispersal (Rickefs and Renner, 1994).

The dispersal of pollen and seeds, which we have just approached from an empirical point of view, can be treated mathematically through game theory (Maynard-Smith, 1982), which will not be discussed here because it goes beyond the scope of floral morphology in the strict sense. We limit ourselves to saying that it is possible to study theoretically the strategies optimizing fitness and involving pollen heteromorphism (number of apertures, Till-Bottraud et al., 1994), or seed heteromorphism (e.g., with or without a parachute, in *Crepis* sp., Asteraceae).

5.2. SEXUALITY AND FERTILITY

We will not take up the topic of floral sex determination here, our aim being rather to understand the evolution of floral sexualization in relation to fertilization strategies. Most species of Angiosperms are hermaphrodite or monoecious and therefore can inherently be expected to have a potential for self-fertilization. However, we have seen in the earlier chapters that there are strategies (architectural or molecular) preventing self-fertilization and favouring allogamy. Hence, the prevention of self-fertilization seems to be a general rule, provided a quantitative study of self-fertilization validates the hypothesis.

Things are not so simple, however, because among other things, it is difficult to measure the degree of allogamy in a given progeny. Electrophoreses can be used, through allozymic profiles (Brown and Allard, 1970), to calculate the proportion of seeds produced by self-fertilization, s, and by cross-fertilization, $t = 1 - s$. This technique is applicable only to the female function of the plant and not to all the progeny. Other techniques can be used to obtain this category of parameters, such as the use of microsatellites. Two types of data will thus be the subject of the following sections: measurement of fertility and mode of fertilization (self-fertilization, cross-fertilization, agamospermy[1]).

5.2.1. Fertility and Parameters of Fitness

The study of the reproduction of Angiosperms often concerns the production of seeds by a given plant as an indicator of fertility. However, Angiosperms are mostly hermaphrodites and hence the male contribution to sexual reproduction should be considered, that is, the evolutionary processes have an effect on both male and female reproductive aspects. At the population scale, since

[1]Note that *agamospermy* is very often confused with *apomixis*. Apomixis (from *apo,* without, and *mixis,* union) etymologically means a reproduction without mating of two gametes and thus includes all the processes of asexual (vegetative) reproduction. The formation of seeds without fertilization (parthenogenetic organisms) is agamospermy (*a* privative, *gamos*, marriage, and *sperma*, seed).

each plant has a father and mother, it is supposed that each gender contributes equally to reproduction (Charnov, 1982), so that the fitness of an individual *i* of a population, denoted as $W(i)$, is expressed as the average relative male and female fitness (that is, divided by the average fitness in the population), according to the following equation:

$$W(i) = \frac{1}{2}\left[\frac{W_f(i)}{\overline{W_f}} + \frac{W_m(i)}{\overline{W_m}}\right]$$

That being stated, evolutionary problems can be investigated in terms of variations of W as a function of certain parameters. Mathematically, values called **marginal values** are used (Lloyd, 1988). For example, suppose a floral trait x, capable of varying from one generation to the next; in this case the marginal value $m(x)$ of W relative to x will be the partial derivative of W with respect to x:

$$m(x) = \frac{\partial W}{\partial x} \approx \frac{\Delta W}{\Delta x}$$

For example, we may ask whether female fertility is limited by the accessibility of the stigma for the pollen or even the availability of resources for maturation; studies seem to show that these two aspects limit female fertility (Haig and Westoby, 1988). However, this type of analysis of floral traits (or characters), or of resource allocation, raises a question of interpretation because it is difficult to dissociate some non-independent parameters. For example, a modification of petal length diminishes the accessibility of stamens. Hence, calculations of $m(x)$ cannot be used to understand what characters are the targets of selection. Moreover, relationships between parameters that are apparently distantly related also need to be considered, such as the establishment of a trade-off between the development of floral characters and fertility (production of gametophytes).

5.2.2. Problems Linked to Self-fertilization

The possibilities of self-fertilization are not negligible, not only because the majority of Angiosperms are hermaphrodite, but also because flowers are often arranged in inflorescences and not isolated. Thus, Angiosperms do not seem to be protected from inbreeding depression.

a) General points

A typology of self-fertilization processes can be drawn on the basis of space or time (Lloyd, 1979). There are several possibilities: (1) self-fertilization "prioritized" over cross-fertilization (the ovules are spontaneously fertilized without an opportunity for allogamy), (2) competition between foreign pollen grains and self-pollen, and (3) delayed self-pollination (self-fertilization occurs when the opportunity for cross-pollination has been lost).

The general term *self-fertilization* comprises two possibilities:

- autogamy in the strict sense, where a stigma is fertilized by pollen from the same flower;
- geitonogamy, where one flower fertilizes another flower of the same inflorescence that is receptive at the same time.

Even though the fitness of a plant, $W(i)$, can be expressed at the outset simply in terms of male and female fertility of flowers ($w_f(i)$ and $w_m(i)$) and of the number of flowers $n(i)$, according to the following formula:

$$W(i) = \frac{1}{2}\left[\frac{n(i)w_f(i)}{\overline{W_f}} + \frac{n(i)w_m(i)}{\overline{W_m}}\right]$$

that does not take into account self-fertilization aspects. The overall fitness is especially affected by self-fertilization since the zygotes resulting from self-fertilization do not have the same fitness (probability of survival and growth) as those resulting from cross-fertilization. Therefore, we cannot simply multiply $n(i)$ by $w(i)$.

Evolutionary processes probably also have an effect on $n(i)$, which is likely to result from a trade-off between the consequences of inbreeding depression and the optimization of the male function. Indeed, plants with large inflorescences tend to export more pollen but are subject to more geitonogamous fertilization. Thus, we can suppose that for each species there is a given inflorescence size, not too large and not too small, representing an equilibrium between these two types of evolutionary constraints (see Chapter 1).

b) Importance of self-fertilization

There is a tendency to believe that Angiosperms demonstrate an extreme optimization of cross-fertilization sometimes by highly sophisticated methods of attracting pollinators, etc., in order to avoid self-fertilization at all costs. One argument advanced is that self-fertilization could lead to the formation of individuals homozygous for harmful alleles, which are the source of inbreeding depression. This rather exclusive idea presents a certain number of problems.

1. Suppose that in a population a gene promoting self-fertilization appears. The frequency of this gene will increase because on average the individuals realizing self-fertilization transmit more copies of genes than do allogamous individuals. Indeed, in one of its seeds, an allogamous individual carries one copy of that gene, by means of the ovule, and in a nearby seed, it also carries a single copy by means of a pollen grain. Thus, in an allogamous system, the

average contribution of an individual is two copies of the gene. On the other hand, an autogamous individual carries a copy by means of the ovule to the following generation and either one copy (null self-fertilization) or two copies (self-fertilization at a rate equivalent to that of cross-fertilization) through the pollen, that is, between two and three copies (supposing that self-fertilization does not exclude cross-fertilization). This greater genetic contribution of non-obligatory autogams (or non-obligatory allogams) in relation to obligatory allogams (Fisher, 1941) is called the "automatic" selective advantage of self-fertilization.

This reasoning supposes, moreover, that the pollen stock is not limiting, i.e., that the pollen grains used for self-fertilization are not unavailable for cross-fertilization. Nevertheless, a hypothesis of limitation by the available pollen quantity is not ruled out (Holsinger, 1991).

Under that hypothesis, allogamy is favoured if the fitness of individuals resulting from self-fertilization, w_s, is lower than that of individuals resulting from cross-fertilization, w_o, so that an inbreeding depression is measured as $\delta = 1 - w_s/w_o$. It is possible to calculate that the advantage of self-fertilization described above disappears if $\delta > 0.5$, i.e., when $w_s < 0.5w_o$ (Charlesworth, 1987). It can therefore be supposed that two majority cases (stable states) will present themselves: self-fertilization is greatly widespread (because $w_s > 0.5w_o$) or not widespread (because $w_s < 0.5w_o$) and, much more rarely, there is coexistence (strict equality $w_s = 0.5w_o$). A preliminary validation of these considerations is the measure of the frequency of self-fertilization in plant populations: one study involving 129 species belonging to 67 genera and 33 families shows in fact a bimodal distribution of self-fertilization frequency (Barret and Eckert, 1990).

2. The formation of homozygotes for harmful alleles by self-fertilization allows a "purging" of these alleles from the population. Self-fertilization is perhaps not therefore as harmful as earlier believed. For example, in water hyacinth (*Eichhornia paniculata*, Pontederiaceae), systematic self-fertilization has been repeated over five generations in an initially allogamous population, and in an initially autogamous population. Eventually, an allogamous fertilization was realized. The fitness remains stable in the initially autogamous population, but in the initially allogamous population it decreases to rise again to its original level following the final allogamous fertilization (Barrett and Charlesworth, 1991). It is probable that autogamous populations are slow to express inbreeding depression, which especially explains high rates of δ observed in some cases, unlike in allogamous populations, where the effects of inbreeding depression are rapid. However, a high rate of mutation could partly balance these effects and thus maintain a high value of δ in populations that have only a moderate frequency of self-fertilization.

The two arguments that have just been developed show that non-obligatory self-fertilization is possible and even that it is perhaps not rare in nature. How can we quantify the impact on δ of self-fertilization rates? One method is to consider the inbreeding coefficients of individuals from two successive generations in a population in which self-fertilization is declared at a rate s. The mean inbreeding coefficient of adult individuals at generation n is denoted as F_n and that of the zygotes they form is F_n^*. These zygotes will thus undergo the effects of selection, since those resulting from self-fertilization have a fitness w_s and those resulting from cross-fertilization have a fitness w_o. The inbreeding depression is $\delta = 1 - w_s/w_o$. The inbreeding coefficient of adults at generation $n + 1$ is thus F_{n+1}.

By definition, allogamous individuals n produce homoallelic zygotes with the probability F_n. Thus, the probability of obtaining homoallelics by this means is $(1 - s)F_n$. The probability of obtaining homoallelic zygotes by means of autogamous homoallelic individuals n is $s \cdot F_n$, and the probability of obtaining them by means of autogamous heteroallelic individuals n is $s \cdot (1 - F_n)/2$. Hence:

$$F_n^* = (1-s)F_n + \frac{s(1-F_n)}{2} + sF_n = (1-s)F_n + s\frac{1+F_n}{2}$$

We define here:

$$G_n^* = (1-s)F_n \text{ and } H_n^* = s\frac{1+F_n}{2}$$

Four groups of zygotes can be formed: those resulting from self-fertilization and homoallelic or heteroallelic, those resulting from cross-fertilization and homoallelic or heteroallelic. Their respective frequencies are H_n^*, $s - H_n^*$, G_n^*, $(1 - s) - G_n^*$, and their selective values are respectively w_s, w_s, w_o, w_o. After selection, the frequencies of these zygotes are:

$$\frac{w_s H_n^*}{\Gamma}, \frac{w_s(s-H_n^*)}{\Gamma}, \frac{w_o G_n^*}{\Gamma}, \frac{w_o(1-s)-G_n^*}{\Gamma}, \text{ where } \Gamma = sw_s + (1-s)\,w_0$$

Thus:

$$F_{n+1} = \frac{w_s s\left(\frac{1+F_n}{2}\right)}{\Gamma} + \frac{w_o(1-s)F_n}{\Gamma} = \frac{s\frac{w_s}{w_o}\frac{1+F_n}{2} + (1-s)F_n}{s\frac{w_s}{w_o} + (1-s)}$$

As $\delta = 1 - w_s/w_o$, we get the following:

$$\delta = \frac{F_n^* - F_{n+1}}{H_n^* - sF_{n+1}} \approx \frac{F_n^* - F_{n+1}}{F_n^* - sF_{n+1}} \text{ if } F_n^* \approx H_n^*$$

It is thus seen that the depression δ comprises at the numerator the deviation of inbreeding coefficients between zygotes and adults formed. If F is represented over time (discrete since it is counted by means of n generations), we will observe a saw-tooth graph: F will increase from F_n to F_n^* under the effect of s and then diminish from F_n^* to F_{n+1} under the effect of δ (Fig. 5.1). Moreover, on the basis of the expression of F_{n+1} as a function of s explained above we can write the following:

$$\Delta F_n = F_{n+1} - F_n = \frac{s(1-\delta)\left(\frac{1-F_n}{2}\right)}{1-s\delta}$$

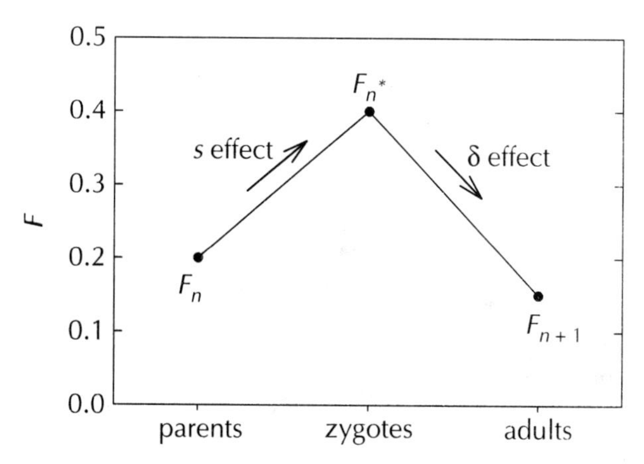

Fig. 5.1: Diagram of the inbreeding coefficient F as a function of the generation (see text) (modified from Barrett et al., 1996).

For a fixed F_n, we see that the deviation of the inbreeding coefficient of adults between generation n and $n+1$ is lower than if it were proportionate to s, because of the denominator. For a high value of δ, this deviation is relatively low, which is related to the "stability" of the fitness observed further up in the autogamous populations of water hyacinth: the purge effected by selection (low w_s) compensates the expression of harmful alleles promoted by self-fertilization.

c) Self-fertilization and pollen

The study of the processes of autogamous and geitonogamous self-fertilization also involves the distribution of pollen grains. Indeed, when a pollinator visits several flowers of a single inflorescence (which often happens, especially in the study model, water hyacinth), it transfers to a given flower, say number j, a certain quantity of pollen grains, which are distributed into S_j grains deposited on the stigma of the same flower and T_j grains remaining attached to it. The quantity $Q_j = T_j/(S_j + T_j)$ gives the proportion of grains exported by flower number j. When the insect visits the subsequent flowers, a proportion q is deposited and $(1 - q)$ remains attached to its body. At the end of $k - 1$ visits, the pollinator arrives on the kth flower, has an additional proportion $Q_j(1 - q)^{k-1}$ of grains attached to its body, and delivers out of that the proportion $qQ_j(1 - q)^{k-1}$. In total, the pollinator has deposited:

$$D = \sum_{j=1}^{n-1}\left(Q_j \sum_{k=j+1}^{n} q(1-q)^{k-j-1}\right) = \sum_{j=1}^{n-1} Q_j(1-(1-q)^{n-j})$$

It is easy to demonstrate that:

$$D \xrightarrow[n\to\infty]{} \sum_{j=1}^{\infty} Q_j$$

if this sum converges. In other words, for a large number of flowers, almost all the pollen exported from flowers will be deposited on nearby flowers. Of course, the number of flowers is not infinite in reality, and we must therefore calculate D. What is important to remember is that D increases with n, the number of flowers visited. That only repeats the intuitive result that an increase in inflorescence size increases the possibilities of geitonogamy, and shows also very clearly that, for sufficiently large n, the pollen exported from flowers will be used only for geitonogamous self-fertilization, at the cost of cross-fertilization. A way to limit this phenomenon is to decrease the number of flowers per inflorescence (finite n) and/or to change the floral architecture so that Q_j is minimized.

The problems that we have mentioned occur in species that have a stock of pollen that can be separated into grains that are distributed in one way or another. In Orchidaceae (and Asclepiadaceae), the pollen is agglomerated into pollinia; that is why, *a priori*, these plants can be supposed to be particularly subject to geitonogamy. Still, observations made in *Orchis mascula* and *Orchis morio* (Johnson and Nilsson, 1999) seem to show that this is not the case. These two species are called "deceptive" because they do not produce nectar.[2] The pollinators are most often insects of the genus *Bombus*. And indeed, the pollinia extracted from a flower and carried by these insects are not transferred automatically to the stigmas of the subsequent flowers belonging to the same inflorescence. The major reason for this is that a pollinium can be

[2]Most deceptive Angiosperms belong to Orchidaceae (it is estimated that there are 8000 deceptive species of Orchidaceae among the 19,500 species of Orchidaceae in the world).

deposited on a stigma only from a given angle. When an insect shakes an orchid flower while searching for nectar, it often happens to take an adequate position when it searches long enough (from 30 to 80 seconds, depending on the species). In these two species, the insects spend only 30 seconds visiting an inflorescence, probably because there is no nectar, since the artificial addition of nectar causes an increase in the visit time per flower, and in the rate of geitonogamy. All other things being equal, a strategy such as "deceit", or more generally avoidance of geitonogamy, seems to be possible only when pollinators are present in abundance. But this is not the case, at least with respect to these two species, since it can be shown that the quantity of pollinators is a limiting factor of seed production. It is possible that the absence of nectar in some Orchidaceae results from a trade-off between attraction, i.e., development of perianth (and fragrance), and nectar production. In nectar-producing Orchidaceae such as *Platanthera bifolia*, the time required before a suitable positioning is 80 seconds, i.e., very long.

d) Structure of populations

From all these considerations about self-fertilization that we have just discussed, some hypotheses can be formulated about the structure of plant populations. Because flowers are arranged generally in inflorescences and plants multiply also by asexual reproduction, and because the floral traits promote self-fertilization to a certain extent, it can be supposed that a plant population is made up of patches that have a low to medium level of inter-patch inbreeding within which the individuals are similar (high F). Thus, this hypothesis considers a population as a mosaic of sub-populations in which self-fertilization is high, cross-fertilization between different sub-populations allowing the diffusion of alleles on the scale of the total population (Barrett et al., 1996). Moreover, genetic studies tend to show that the levels of inbreeding of different lines of plants are not equal to the level of the population overall (Mutikainen et al., 1998), especially because each line has a given history, in terms of self-fertilization, so that it shows a particular value of δ.

In this context, the role of pollinators must be emphasized because it achieves this allelic "diffusion", and because the particular behaviours of varied pollinators will have an effect on that diffusion: some insects such as bumblebees tend to pollinate within a single sub-population, while others such as butterflies travel long distances between two successive flowers, mixing the alleles between sub-populations.

5.2.3. Agamospermy and Parthenogenesis

In some genera, agamospermy (or what some wrongly call **apomixis**) has nearly replaced sexual reproduction, which is observed in most other species.

This is especially the case with lady's mantle (*Alchemilla* sp., Rosaceae). In *A. vulgaris*, pollen production is greatly altered because most of the tetrads degenerate. However, cases of obligatory agamospermy are much rarer than those of facultative agamospermy, as in the hawkweeds (*Pilosella* sp., Asteraceae). There is thus formation of "apomictic" embryos without fertilization; these could accompany embryos resulting from fertilization and the two types coexist, even within a single seed. This is especially the case with facultative agamospermy in *Citrus* (Rutaceae), which is the source of what are called **adventitious** embryos. Following fertilization of the egg and the central cell, the embryo (proper and endosperm) develops, apparently activating the production of adventitious embryos from cells of the nucellus. These embryos compete with the principal (sexually formed) embryo—and between each other, when there are several in the same seed—and in favourable cases completely supplants it. In other cases, the adventitious embryo and the sexually formed embryo coexist and share the same endosperm.

> From a certain point of view, agamospermy seems to be a completely asexual process since it is a cell of the nucellus that produces an embryo. However, parthenogenesis (from the Greek *parthenos*, virgin) is more generally (in the angiosperms) the production of sporophytic individuals in the form of seeds from cells of the germinal line without fertilization. Strictly speaking, it is not agamospermy as we have described it, but rather the production of an embryo from the egg without fertilization (there are two possibilities: either the egg has undergone a chromosomal reduction followed ultimately by a duplication that re-establishes $2n$ chromosomes, or the egg has not undergone chromosomal reduction). Thus, there is a certain terminological flaw: while agamospermy and parthenogenesis seem synonymous, they are not so, because the first is most often reserved for the nucellar case, and the second for cases of germinal cells. We will refer to the two processes together as parthenogenesis *s.l.*

In any case, parthenogenesis *s.l.* is not negligible and reduces the quantity of egg cells able to form individuals by fertilization by a pollen grain. In terms of seed formation this does not significantly decrease the reproductive success of the plant (and could increase it in some cases). In *Rubus* sp., apomictic populations have a restriction polymorphism that is weaker than that of populations that reproduce sexually. The restriction profiles of seeds produced by obligatory apomictic lines in dandelion (*Taraxacum* sp., Asteraceae) are apparently hardly variable, except those resulting from some rare lines in which the appearance of a mutation is supposed. Other studies on dandelion also seem to show that the apomictic lines have levels of heterozygosity and restriction polymorphism as high as in non-apomictic allogams, and much higher than in self-fertilizing lines. The impact of parthenogenesis remains to be defined, therefore, since it does not appear to be equivalent to self-fertilization in genetic terms (for review see Briggs and Walters, 1997).

5.3. POLLEN DISPERSAL AND FERTILIZATION IN HETEROGENEOUS CONDITIONS

The heterogeneity of pollen dispersal is apparent on at least two levels: in the flower and during pollen transport. The non-homogeneity of flowers allows the pollinator to "choose" its target, a process that has an effect on the relative fertility of each floral type. In pollen transport, the heterogeneity lies in the fact that the conditions of transport are not constant, notably because the animal pollinator is not always of the same species.

5.3.1. Pollination and Selection

a) General points

The dispersal of pollen by pollinators is a key point in the reproduction of zoophilous Angiosperms. Attraction as well as pollen deposit are processes that could be under selection. Thus, we might at first suppose that the floral architecture is subject to selection pressure linked to the pollinator. However, that does not seem to be obvious because variations in pollinators do not always lead to visible changes in the floral structure. In *Polemonium* sp. (Polemoniaceae), a change in the corolla size following an artificial change in pollinator was observed in just a few generations (Lloyd et al., 1996). A widely studied example of floral selection by insects involves colour. Indeed, the polymorphism of the corolla colour originates from mutations affecting pigment synthesis proper or its control. White flowers are a well-known case of chromatic polymorphism that has been studied in foxglove (*Digitalis purpurea*, Plantaginaceae[3]), larkspur (*Delphinium nelsonii*, Ranunculaceae), and morning glory (*Ipomea purpurea*, Convolvulaceae). Discrimination between chromatic variants is expressed as a reduction in seed production and pollen export, and by a higher rate of self-fertilization than in the wild type (Levin and Brack, 1995). In larkspur especially, individuals with white flowers produce fewer seeds than those with coloured flowers. The insects visit white flowers less often because the development of their nectaries is slower.

It seems clear that pollinators such as bumblebees (*Bombus* sp.) have a preference for the most common coloured forms (or **morphs**) of flowers in a given species, a process that gives rise to **positive frequency-dependent** selection giving an advantage to the most common morph. However, in some

[3]Formerly Scrophulariaceae.

deceptive Orchidaceae (see above) such as *Dactylorhyza sambucina* or *Orchis caspia*, all other things being equal (especially with respect to nectar), pollinators prefer the rare morphs. This is a **negative frequency-dependent** selection, promoting polymorphism of the corolla colour (Smithson et al., 1997).

Generally, the consequences of selective processes running between pollinators and floral structure are poorly understood, because they vary from one taxon to another and are difficult to quantify. However, precise studies have occasionally been carried out, for example involving pollination by hummingbirds, which are taken as an example here (Melendez-Ackerman et al., 1997; Cambell et al., 1996; Murcia and Feinsinger, 1996).

b) Ornithophilous pollination

In flowers pollinated by hummingbirds, selection occurs at three levels: colour, size of the perianth, and floral architecture as a whole.

Floral architecture

Pollinators often visit several species of plants in a single phase to search for food, which leads to interspecific pollen transfers (IPTs), reducing the reproductive success of the individuals considered. For example, heterospecific pollen grains can considerably disturb the germination of homospecific grains. Moreover, pollen grains of a given individual could be lost during a visit to a heterospecific flower by friction or other causes. The floral architecture becomes important from this perspective, because the precise places where the pollen is collected or deposited by the pollinator depend on the plan of floral organization. Hence, flowers having very different architecture will undergo few IPTs. The theory that sympatric plants having common pollinators will diverge structurally (particularly in the arrangement of sexual organs) is referred to as the **sexual architecture hypothesis** (**SAH**).

Hummingbirds do not discriminate between species and are probably responsible for a large number of IPTs. Studies have been conducted in the Costa Rican forest on a self-compatible shrub, *Palicourea lasiorrachis* (Rubiaceae) and its almost exclusive pollinator, the hummingbird *Lampornis calolaema* (Murcia and Feinsinger, 1996). There are five other species of bushes that flower at the same time and share the same pollinator: *Cephaelis elata* (Rubiaceae), *Hansteinia blepharorachis* and *Dicliptera iopus* (Acanthaceae), *Besleria triflora* (Gesneriaceae), and *Satyria warszewiczii* (Ericaceae). Observations confirm that a single visit to a heterospecific flower is enough to decrease considerably the initial pollen load of *P. lasiorrachis*, probably not because of the deposit of pollen grains of *P. lasiorrachis* on the stigma of subsequent flowers but rather because of a scraping by non-sexual parts (especially corolla). The IPTs here reduce the quantity of pollen available to the subsequent *P. lasiorrachis* flower by around 75%. Also, the scraping is more effective when there are species with a floral organization plan that is greatly different from *P. lasiorrachis*: *S. warszewiczii, H. blepharorachis*, and *D. iopus*.

Flower size
So far we have looked at the quality of the pollen load carried by the pollinator. The floral architecture also controls the quantity of pollen exported, particularly by means of accessibility to the stock of pollen. One parameter determining this accessibility is flower size. In *Ipomopsis aggregata* (Polemoniaceae), individuals with larger flowers export more pollen (Campbell et al., 1996). Even though larger flowers certainly produce more pollen, which could explain the greater pollen export, this export is also strongly promoted by deeper penetration of the hummingbird's beak. It has also been reported that there is a selection in favour of a larger flower size, thus creating a sort of structural "correspondence" or fit between the plant and the pollinator.

Flower colour
Studies of chromatic preferences of pollinators sometimes use flowers with different architecture because there is effectively a link between structure, taxon, and colour. In these conditions, it remains difficult to draw conclusions because colour, nectar production, and other parameters are not independent. In natural populations of *Ipomopsis* sp., the pollinators (hummingbirds of the genus *Selasphorus*) seem to prefer the red flowers of *I. aggregata* to the white and narrower flowers of *I. tenuituba*, very likely because of a lower nectar production in the latter. Moreover, the hummingbirds prefer (1) white flowers when they contain more nectar, as shown in experiments that added nectar artificially, or flowers of *I. aggregata* when the flowers of *I. tenuituba* are stained red, (2) red flowers when some flowers of *I. aggregata* are bleached white or some flowers of *I. tenuituba* are stained red. Thus, although a chromatic preference of hummingbirds is shown, and although generally a chromatic preference of pollinators can be supposed, it is important to place these considerations in an ecological context: the selection operated by pollinators can be broken down into independent components (colour, form, and nectar), but these components are not independent from the point of view of the flower.

5.3.2. Pollen Transport

Pollen transport has long been the focus of investigation (studies of Bateman et al.), since it constitutes a key point of the reproduction of Angiosperms with biotic pollen dispersal, because variations of transport conditions have repercussions on the fertility of the plants considered. There are three levels of heterogeneity in pollen transport:

— "Horizontal" heterogeneity: the pollen is deposited on variable places on the body of the pollinator (which is assumed to be an insect), each of them having different probabilities of pollen loss (by falling) or deposit on a stigma.

—"Vertical" heterogeneity: pollen collected during successive visits of flowers by the insect accumulates in layers on the body of the insect; from that point, the pollen on the deeper layers is no longer directly accessible to the stigmas.

—The insect is not always of the same species; the efficiencies of extraction and abandonment of pollen are thus variable.

a) Horizontal heterogeneity

The most simple mathematical formulation of floral export of pollen considers two parameters only: π, the probability of "non-fall" of pollen that is to be collected, and ρ, the probability of pollen deposit on a stigma. The ith flower visited receives from the flower initially visited the quantity D_i of pollen, according to the following formula:

$$D_i = \pi R \rho (1-\rho)^{i-1}$$

where R is the quantity of pollen removed from the initial flower. Two aspects can be added to refine this model: a possibility of pollen loss from the body of the insect and the existence of two types of storage sites on the body of the insect, one secure, the other exposed, i.e., from which the pollen can fall (Bertin, 1997). The pollen on the body of the insect is redistributed when the insect grooms itself. Four stocks of pollen are thus distinguished: R, S, E, and D. The quantity of pollen received by flower number i is written as follows:

$$D_i = \rho_S (S_{i-1} + \Gamma \gamma_S E_{i-1}) + \rho_E (1-\Gamma) E_{i-1}$$

where S_{i-1} and E_{i-1} are the secure and exposed stocks of pollen following the visit to flower number $i-1$. Γ is the total quantity of pollen leaving the exposed site, by falling (probability γ_L) or by slipping towards the secure site (probability $\gamma_S = 1-\gamma_L$). From the systems of recurrence equations (for each stock), we can ultimately write the general expression of D_i as a function of i:

$$D_i = R(\omega \chi^{i-1} + \varphi \xi^{i-1})$$

where:

$$\omega = \rho_S \left(\pi_S + \frac{\pi_c \Gamma r_s (1-\rho_S)}{\rho_E + \Gamma(1-\rho_E) - \rho_S} \right) \text{ and } \chi = 1-\rho_S$$

$$\varphi = \pi_E (1-\Gamma) \left[\rho_E - \frac{\Gamma \gamma_S \rho_S (1-\rho_E)}{\rho_E + \Gamma(1-\rho_E) - \rho_S} \right] \text{ and } \xi = (1-\Gamma)(1-\rho_E)$$

The form obtained resembles the first intuitive equation with two parameters, to which can be added a corrective term ξ. In particular, it is noted that when γ_L is high, i.e., when the probability of loss from the exposed sites is high, the term ω declines in favour of the term φ, diminishing D_i considerably because ξ is the product of quantities smaller than 1.

b) Vertical heterogeneity

Here it is supposed that pollen accumulates in three distinct layers: surface (T), middle (U), and lower (L). Only the grains of the surface layer can be deposited on a stigma. It is supposed that the lower layer is a dead end for the pollen grains: once they are in that layer, they will always remain there. The probability of sinking into a lower layer is μ. This is corrected by a multiplication factor ξ for the transition from U to L. The probability of a return from U to T or from T to D is ρ. The system of recurrence equations is the following:

$$T_i = (1-\rho)(1-\mu)\,T_{i-1} + \rho(1-\mu)\,U_{i-1}$$

$$U_i = (1-\rho)\mu\,T_{i-1} + [(1-\rho)(1-\xi\mu) + \mu\rho]U_{i-1}$$

$$D_i = \rho\,T_{i-1}$$

The resolution of the system ends in a general expression of D_i, of the following type:

$$D_i = \frac{R\rho\pi_r}{2K_0}(K_1\lambda_1^{i-1} + K_2\lambda_2^{i-2})$$

c) Heterogeneity linked to type of pollinator

It is possible in the earlier cases to reason in terms of flux between compartments with probabilities of transfer, but that is no longer the case with heterogeneity linked to type of pollinator because all the species that could pollinate a given flower cannot be known. This is why a probabilistic reasoning is adopted using expectations of (random) variables D_i. It is supposed that the probability of initial loss depends on the floral structure alone and is thus fixed at π. On the other hand, the parameter ρ depends on the insect, and it is supposed that it follows a normal (Gaussian) truncated law ($\rho \in [0; 1]$). The quantity that is received by flower number i is what remains after visit $i-1$, minus what remains after visit i, which is expressed as follows:

$$E(D_i) = \pi R\,[E((1-\rho)^{i-1}) - E((1-\rho)^i)]$$

where E is the expectation function.

In nature, the pollen dispersal encompasses the three aspects of the preceding sections a, b, and c and is mathematically expressed by a combination of solutions. Moreover, in all the equations, the parameter R has been used as being fixed; however, this parameter is a function of the production of pollen, which depends on the time, anther position, and degree of opening of pollen locules.

d) Consequences on floral morphology

A priori, the parameter that is important for the initial flower is the quantity of pollen that the insect can carry and that will be effectively dispersed on the stigmas. Moreover, according to the reasoning that we have adopted, it seems that the male stochasticity (on R, π, ...) affects the entire dispersal chain, which is not the case with the female, in which the effects of the variability of ρ progressively increase as the number of visits increases. This is so because the problems are different. With a given pollinator, the initial flower has one and only one opportunity to distribute pollen, whereas there are n stigmas each having an opportunity to receive pollen. If i is large enough, from the initial flower there remains nearly nothing at flower number i. By means of a similar calculation to that carried out in section 5.2.2c, we can show that if n approaches $+\infty$, the quantity of pollen no longer depends on ρ. For example, using a model of dispersal in **homogeneous** conditions (i.e., Bateman model), the quantity distributed at the end of the flower n would be W_n:

$$W_n = \sum_{i=1}^{n} D_i = \pi R(1-1-\rho)^n) \xrightarrow[n\to\infty]{} \pi R$$

the limit of which does not depend on ρ.

However, this is valid only in a homogeneous model. Indeed, the limit of W_n in heterogeneous conditions is still a function of ρ (and other parameters of the model). Once again, in order to set our ideas down, in the case of a model with horizontal heterogeneity, the limit of W_n when n tends toward $+\infty$ is the following:

$$W = R(\pi_s + \varepsilon\pi_E) + \frac{R\pi_E(1-\Gamma)\,(\rho_E - \varepsilon'\rho_s)}{\Gamma + \rho_E(1-\Gamma)}$$

where ε and ε' are two terms expressed as a function of Γ, γ_S, and ρ, which are considered here as corrective multipliers for π and ρ respectively. Thus, in the case of a dispersal in heterogeneous conditions, it can be predicted that the optimization of parameters π and R alone will not suffice to optimize the pollen dispersal. One procedure to get the triplets (π_S, π_E, ρ_S) that maximize the dispersal is to calculate the proportion P of pollen removed from the initial flower and dispersed on flowers number i, $i > k$, k being a fixed threshold. This threshold k is imposed in relation to the self-incompatibility between flowers of a single inflorescence: it is supposed that more than k visits are required for the deposit of pollen to be effective, i.e., for the insect to finally leave the group of flowers that are incompatible.

In the case of horizontal heterogeneity (i.e., of insects that groom themselves), it is observed that *P* is maximal for high values of π_t (= $\pi_S + \pi_E$) and low values of ρ_S (Fig. 5.2A). In the case of vertical heterogeneity (where the pollen accumulates in layers), an equitable contribution of male parts (π_t) and female parts (ρ) maximizes *P* (Fig. 5.2B). From an anatomical point of view, it means that in the first case there is a tendency towards a significant development of the androecium in favour of dispersal; in the second case the mechanism is less clear because it is not possible to assume simultaneously a very high male and female development since, for a given π_t, *P* is maximal for an intermediate (and not maximal) ρ, and since there is a trade-off between the two.

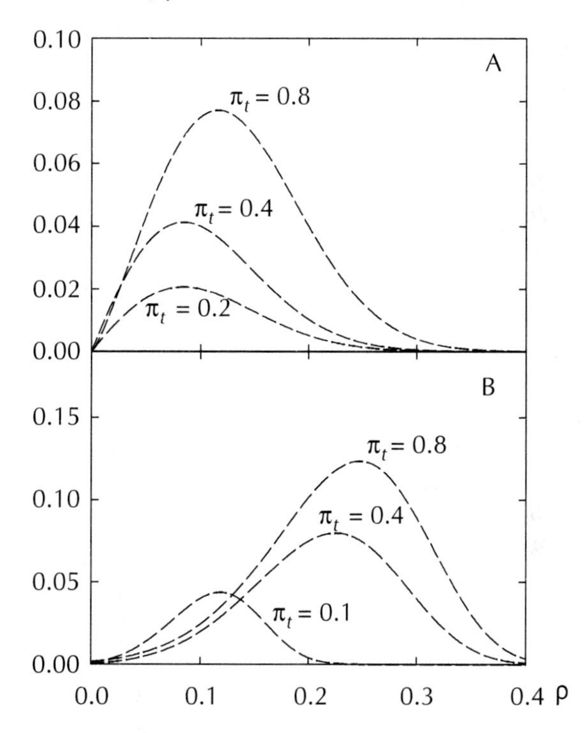

Fig. 5.2: Curves giving the fraction of pollen removed and then dispersed on flowers of index higher than 8, as a function of ρ, in the case of horizontal heterogeneity (A) and vertical heterogeneity (B).

5.3.3. Reproductive Success and Size

a) General points

The attraction of pollinators and the reception of pollen on the stigmas, i.e., the phenomena involved in the "arrival" of pollen on the flower, have greatly interested biologists, but that does not mean that the question of the "departure" of pollen is less important. As we have said, many studies have focused on female reproductive success because it is easy to count seeds, while the tracing of pollen grains and thus of male reproductive success is much more difficult. The male reproductive success (f_m) in nature was at first estimated primarily through methods such as measurement of the rate of visit by pollinators or the export of fluorescent powder applied previously on the anthers. Still, these methods are very indirect and there is not always an obvious relationship between these measurements and f_m. Moreover, f_m also includes problems of pollen germination, i.e., competition between pollen grains. Thus, more recent methods of estimation of f_m use genetic or enzyme markers (Harder and Barrett, 1995).

b) Success curves and floral sexualization

Male reproductive success depends indirectly on female reproductive success, because there is a trade-off between the relative quantities of resources allocated to the male and female functions. This dependence is not always determined with precision, like the curve of evolution of total fitness (f_t) as a function of the proportion of resources allocated to the male function. In particular, f_t may or may not remain constant for any couple (f_m, f_f). The "success curve" consists of graphs representing fitness as a function of allocated resources. The success curve of f_t is horizontal (constant f_t) when there is a population of individuals of identical sizes. When the size of individuals varies, f_t is no longer constant because the increase in size could benefit male function more than female function. The individuals thus adjust the allocation of resources as a function of size: this is what gives success curves their shape (Fig. 5.3).

Size could directly affect the fitness f_m: for example, the pollen of one flower on a tree that is dispersed over a large area and thus undoubtedly has a higher reproductive success than the pollen of a herbaceous plant of the undergrowth. For a similar investment (absolute quantity of resources allocated to reproduction), the f_m of two plants of different sizes is not equal. Further, some large plants will for example produce the same number of flowers as smaller plants and thus possess the same f_m, but if the absolute quantity of resources allocated to the establishment of flowers is identical, the proportion of these resources in relation to the total available resources is smaller in the large plant. With equal proportions (invested in reproduction), a large plant will have a higher fitness f_m.[4] That is why size has an effect on success curves and why this budget effect causes a distortion of success curves that could be at first supposed to be linear.

[4]This type of phenomenon is called the *budget effect*.

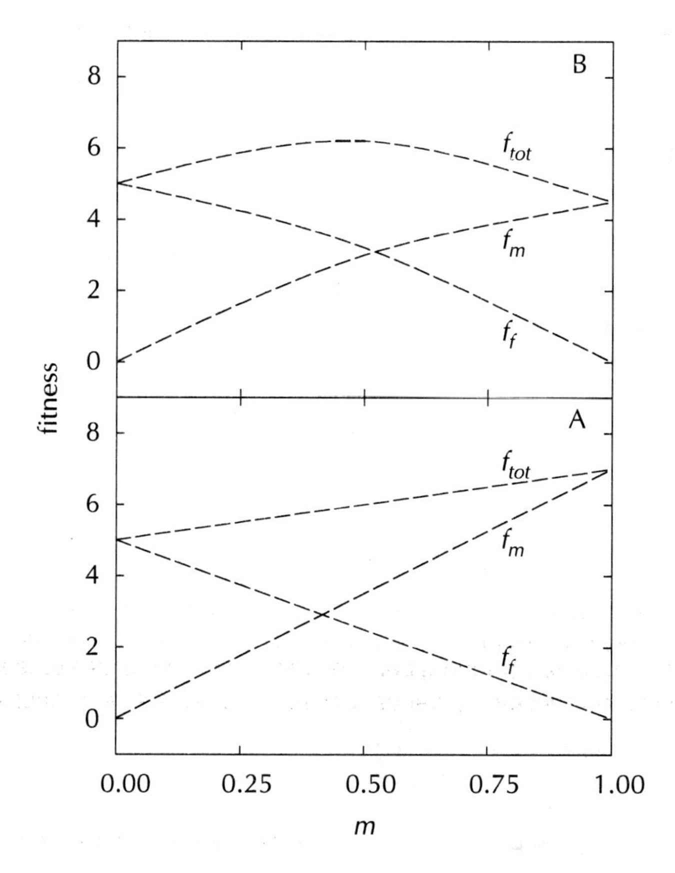

Fig. 5.3: Two examples of success curves representing fitness (f_m, male fitness; f_f, female fitness; f_t, total fitness) as a function of the proportion of resources allocated to the male function (*m*). In both cases, there are differences in size in the corresponding population, and the success curves shown apply to the large individuals. In (A), the fitness is maximal when *m* is equal to 1, so that the large individuals are directed towards the male function. The reverse is true for small individuals, which are directed towards the female function. In (B), there is a budget effect, and hence the fitness is no longer linear as a function of *m* (see text). (Modified from Klinkhamer et al., 1997.)

These success curves show us the possibilities that optimize f_t (i.e., the pair (p_m, p_f), the first being the proportion of resources being allocated to the male function and the second the proportion of resources allocated to the female function,[5] and thus the evolutionary stable strategies or ESS (see Chapter 1).[6] From an analytical point of view, the success curves could be expressed as exponential functions. Since $f_t = f_m + f_f$, we have:

[5]Thus here $p_f = 1 - p_m$.

[6]This expression originates from studies in game theory. The maximization of f_t is a necessary but not sufficient condition for ESS, for which the product $f_m \cdot f_f$ is also maximal.

$$f_t = b_1(Ra)^u + b_2((1 - a)R)^v$$

where b_1 and b_2 are two parameters that may depend on the size *s* (direct effect of *s* on *f*, see above) or may not, in which case the budget effect is such that $u \neq v$; *a* is the quantity p_m. f_t is maximal if *a* annuls its derivative ($a = a_{max}$). Depending on the values of parameters *u* and *v* especially, the allocation (represented by a_{max}) will be to the benefit of the male or female function: the budget effects could affect the male or female function differently and this will occur as a function of the mode of pollination.

Table 5.1 (Klinkhamer, de Jong and Mertz, 1997) summarizes the principal direct and budget effects of size on f_m and f_f.

Table 5.1. Major direct and budget effects of size on male and female fitness. ↑, increase; ↓, decrease

	Zoophilous plants	Anemophilous plants
Direct effects on f_m	Minor and ↓	Major, ↑ or ↓: better pollen dispersal in large plants, but more pollen received by the plant itself
Direct effects on f_f	Variable importance and ↑: better seed dispersal	Variable importance and ↑: better seed dispersal
Budget effects on f_m	Major and ↓: increase of geitonogamy; could lead to pollen saturation of pollinator	Minor and ↓: increase of competition
Budget effects on f_f	Major and ↓: local increase of competition for resources	Major and ↓: local increase of competition for resources

With a given species, the success curves can be used to calculate a_{max} and thus find the proportions of resources allocated to the male and female function, i.e., the adjustment of the gender to size. This would be a primary ecological justification for floral sexualization. However, about 75% of Angiosperms are hermaphrodite and their size could vary from 1 to 100 in a single species: hermaphrodite plants do not seem to verify the theory of sexual allocation, in light of our present understanding. Nevertheless, when the observations are restricted to monocarpic plants, such as *Cynoglossum officinale* (Boraginaceae), the floral gender seems to be linked to size: the small individuals accentuate the male function and produce numerous flowers with a low seed production and the large individuals favour the female function, having fewer flowers with a large number of seeds. Similarly, plants with a sex that evolves over time (monoecious and developing male and female flowers at different times) start life as male and evolve towards being female, while even their size increases (Freeman, Harper and Charnov, 1980).

Note that the budget effects are less significant in anemophilous plants, which implies that their success curves are "more linear" (i.e., closer to straight lines) than those of zoophilous plants. The direct effects of size seem to favour large plants (bushes or trees). Some studies of data comparing species of different sizes are consistent with this prediction, as are studies within a single species. For example, in mercury (*Mercurialis annua*, Euphorbiaceae), an androdioecious plant (i.e., presenting male individuals and monoecious individuals), the male character is more pronounced in larger monoecious individuals.

We have discussed floral sexualization overall, i.e., on the scale of the entire individual: size regulates sex. However, within a single inflorescence, the male or female character may change from one flower to another. In certain hermaphrodite plants, such as columbine (*Aquilegia caerulea*, Ranunculaceae), it has been observed that late flowers produce much fewer fruits than early ones. The same applies to seeds in late flowers, where seeds abort most often before the end of their development. The very poor female reproductive success of late flowers does not originate from a limitation of pollen, because an artificial addition of pollen does not change it (Brunet, 1996). Thus, it seems that late flowers do not have the same aptitude to bring progeny to term, and that is why they seem "specialized" in the male function, i.e., pollen production, while early flowers are specialized in the female function.

Fairly precise data have been obtained on this point particularly in Liliaceae, where it seems clear that variations in reproductive capacity of flowers are linked to a reduction in the number of ovules per flower as one goes distally along an inflorescence. The flower shape also changes. Indeed, the length of the corolla undergoes modifications along the axis of the inflorescence. In *Solanum hirtum* (Solanaceae), an andromonoecious plant (i.e., presenting hermaphrodite and male flowers simultaneously), the hermaphrodite flowers arranged distally on a plant in which pollination is prevented have smaller anthers, stigma, and ovules (Fig. 5.4). By contrast, all the hermaphrodite flowers, whatever their position is, are able to produce fruits. On the other hand, when they can achieve fertilization, the flowers arranged distally are staminate and never produce fruits (Diggle, 1995). The corolla of these flowers is smaller and this size reduction comes only from the presence of fruits on the nearby flowers ("fruits" effect). In contrast, the anthers are always shorter in the distal flowers, whether or not there are fruits. There is thus an **"architectural" effect**. This effect is in addition to the **"fruits" effect** cited above, which refers to the distribution of resources between different parts of the plant. However, the order of magnitude of the effect of early fruits on the development of later fruits at more distal positions is poorly understood.

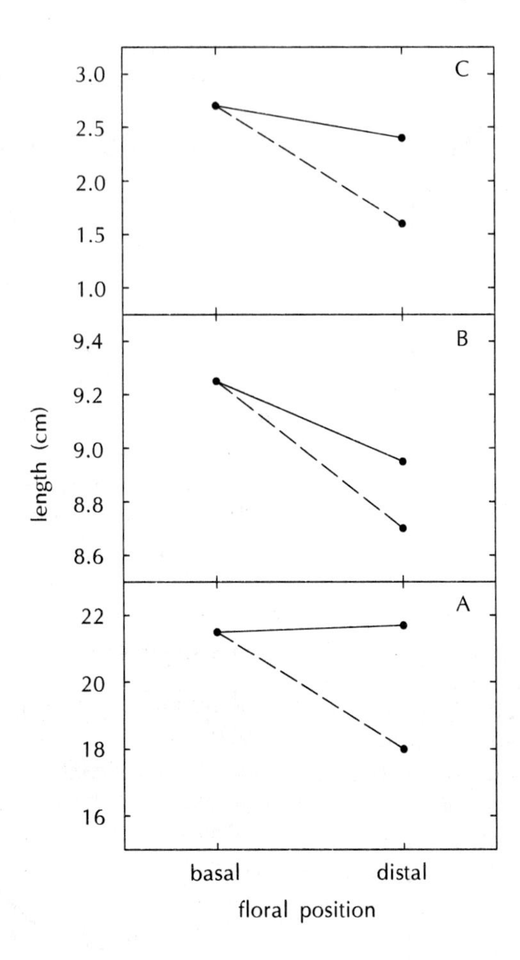

Fig. 5.4. Graph of the length of the corolla (A), anthers (B), and ovary (C) of *Solanum hirtum* as a function of the floral position (distal or basal) in control individuals (solid lines) or individuals that have not been subjected to pollination and have no fruits (broken lines). In (A), only the effect of the treatment ("fruits" effect) is significant, while in (B) only the "architectural" effect is significant. In (C), both effects apply. (Modified from Diggle, 1995.)

c) Other factors influencing sexual allocation

As we have said, it often happens (when the plant is sufficiently large), that there is a bias of resource allocation in favour of female function in bisexual plants. Many reasons have been mentioned, such as self-fertilization (Charnov, 1982), the quantity of pollinators as a limiting factor, competition between pollen grains (Lloyd, 1979), as well as the problem of trade-off between the quantities of resources allocated to reproduction and to vegetative growth

(Burd et al., 1992; Seger et al., 1996). This last point is, however, actively debated. One difficulty of the models proposed lies in the fact that resource allocation is not strictly equivalent in male and female. Indeed, the resources, resulting directly from photosynthesis, are not produced at the same time and, in particular, once the flower wilts, some resources are still allocated to fruit development and thus to the female function.

Continuous models that take these temporal aspects into account show that it is not possible to maximize simultaneously the total resources devoted to reproduction (f_t) and the product $f_f \cdot f_m$ (see note 6 of this chapter), and that the male function can nevertheless dominate in terms of resource allocation, especially if the fructiferous period is short, or that the growth rate during the vegetative period is low (Sakai et al., 1998). Generally, resource allocation seems to be biased if it is impossible to simultaneously optimize f_t and $f_f \cdot f_m$ (the product is maximal for $f_f = f_m$ in the continuous model). The vegetative-reproductive trade-off does not seem ultimately to be the source of the bias toward the female function. The female function starts when other constraints such as those we have mentioned take effect.

5.3.4. Hybridization

Hybridization is an immediate consequence of the proximity of flowers of different species sharing common pollinators. The hybrid individuals, when they exist, often bear intermediate characters. The formation of hybrids could sometimes result in the birth of a new species. This is the theory of **speciation by hybridization**. In this model it is supposed that two non-reduced gametes (in chromosomal terms) of chromosomal number $2n$ and $2n'$ fuse to result in a zygote with $2n'' = 2n + 2n'$ (if the gametes are reduced, a given chromosome of the hybrid inherently cannot have its homologue during meiosis and unless there is one endoduplication, this hybrid will generate no descendants). If the supposedly intermediate characters of the hybrid give it an aptitude for developing in intermediate ecological niches (in terms of resources, thermal or hydric conditions), it could grow without competing heavily with the species from which it was generated. Moreover, a pollen grain from the hybrid will produce a viable zygote only if it encounters an egg that is also a result of the hybrid or of an equivalent hybrid. Otherwise, the zygote formed is effectively aneuploid. Thus, a hybrid population is isolated in reproductive terms.

It is nevertheless possible to envisage the formation of hybrids without assuming non-reduced gametes. This is true especially when the mother species are similar and there are similarities of chromosomal organization, or even when the hybrid zygote undergoes one endoduplication. In any case, the **fertility** of hybrids (F1) remains a problem because, although the hybrids can grow, they are often sterile, at least partly.

Hybridization is not a rare process; a large proportion of species of Angiosperms are supposed to have resulted from it (although that remains difficult to quantify). Today there are manifest examples of hybridization in Orchidae or Rosaceae. The latter family serves here as an example to illustrate the species complexes of the genus *Geum* (Fig. 5.5).

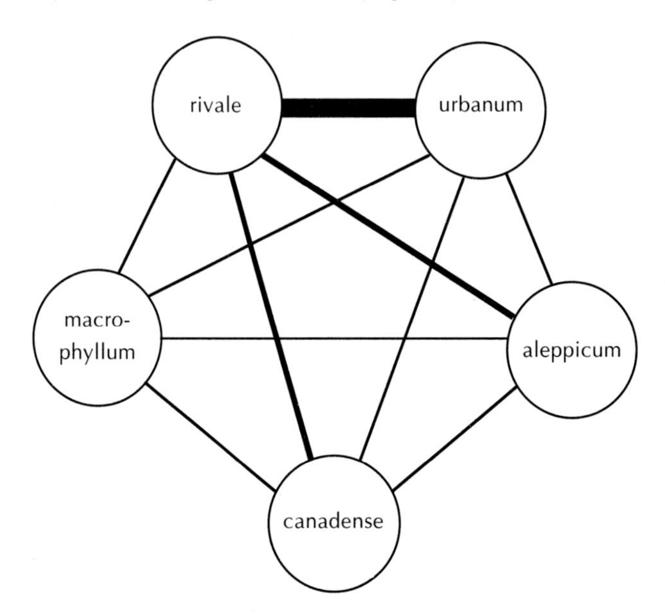

Fig. 5.5. Diagram of modalities of hybridization between some species of the genus *Geum* (Rosaceae). In heavy bold lines, fertile hybrids. In bold lines, partly fertile hybrids. In fine lines, sterile hybrids. (Modified from Briggs et al., 1997.)

Generally, the entire genus *Geum* seems to be susceptible to giving hybrids. The two most widespread species in Europe are *G. urbanum* and *G. rivale*. They are morphologically very different. *Geum rivale* is typically pollinated by Apidae insects (the most common pollinators are of the genus *Bombus*) and has pendant flowers coloured from beige to purple. On the other hand, the erect, small, bright yellow flowers of *G. urbanum* are not specialized and are probably subject to self-fertilization. These two species are frequently sympatric even though there are some differences in distribution: *G. rivale* grows in relatively humid, cool zones, up to altitudes of around 2100 m, while *G. urbanum* grows in shady undergrowth or in areas disturbed by human activities and ordinarily does not grow at altitudes beyond 1800 m.

From the end of the 18th century, the botanist Ehrhart identified a plant he named *G. intermedium*, with intermediate characters between the two species, and recognized it as a hybrid. The mendelian heredity of certain characters as well as the variations of the fruit form in *G. intermedium* were the focus of studies in the first half of the 20th century and even later (Marsden-Jones, 1930; Gajewski, 1957). The probability of hybrid formation seems to vary greatly from one site to another, even when the two parent species are present simultaneously. This probability is always quite low, even during assays of artificial hybridization. On the other hand, the F1 hybrids produced are fertile and the F2 generation shows an entire series of lines with segregated "parental" characters, which suggests the existence of numerous independent genes implicated in the morphological differences. Nevertheless, the formation of *G. intermedium* happens rarely, for the following reasons:

— the pollinators are not the same,
— the flowering periods do not exactly coincide, and
— the sites occupied are different.

The great variability of the presence of hybrids from one site to another is probably explained by environmental effects such as climate: in England, for example, flowering periods seem to overlap greatly. Moreover, the species are close, because the regions degraded by human activities allow the development of *G. urbanum* in places where only *G. rivale* was widespread.

The existence of hybrids implies that these two species are still not entirely separated and that they probably come from a common ancestor. It is tempting to propose that *G. rivale* comes from a "specialization" of a line of *G. urbanum*, following a geographic or ecological isolation. The disturbance of this isolation by man would have led to a combination of the two species.

Even though *G. intermedium* is a result of relatively artificial conditions (generated by disturbances of anthropic origin), this "species" reflects the situation of the genus *Geum*, in which many hybrids have been obtained or observed. The karyotype is hexaploid in *Geum* sp., suggesting that these species have themselves arisen by two successive hybridizations. The cascade hybridizations do not constitute a rare fact, which is why it is supposed, as we have said, that they are important events in the speciation of Angiosperms.

5.4. CONCLUSION

At each step of the process of sexual reproduction, it is possible to construct a model that attempts to reflect reality accurately and allows the drawing of conclusions about the relationships that exist between fitness, floral or inflorescence structure, size, and various other parameters. There are two possible approaches: one genetic, which takes inbreeding problems into account, and the other ecological, which considers the traits (or characters) and attempts to determine the values of traits that optimize the reproductive processes. The first approach tells us that the Angiosperm populations are undoubtedly structured spatially and that self-fertilization is probably not as rare as one might believe, which probably has to do with its function of "purging" harmful alleles. The second approach shows that the floral characters of Angiosperms are the result of complex selective processes that integrate all the stages of pollination and fertilization: parameters such as the number of flowers per inflorescence, floral sexualization, or predominance of a fertile whorl in relation to that of the opposite sex are notably results of an evolutionary compromise or trade-off. Finally, it should be stressed that peculiarities of pollinators (such as grooming behaviour or chromatic "preference") must be taken into account, because these determine the form of the model used. There is clearly a co-evolution between plant and pollinator, since the evolution of floristic characters is, as we have seen, coupled to that of the traits of the pollinator.

Conclusion

Sexual reproduction of living organisms is a process that is often tightly controlled. In animals, there are usually several levels of control:

— Behavioural: The female may choose a sexual partner, based on various traits such as the exuberance of plumage in birds. This phenomenon is part of "mate selection".
— Physiological: For example, because of an endogenous control, hermaphrodite organisms do not have simultaneously male and female potentialities. There is either protandry or protogyny.
— Anatomical: The copulatory organ of certain male insects can remove from the female the semen of the male that previously coupled with her.
— Cellular and molecular: The sexual pheromones of insects, fertilisins in sea-urchins and bindin receptors are examples of cellular and molecular controls.

In animals, it is often said that the female chooses a mate, that choice being apparent (visible to the ethologist) or cryptic (spermatic selection). What we have seen in the preceding chapters justifies the conclusion that, by extension, there is also a "choice of mate" in Angiosperms. This choice, the existence of which can be demonstrated experimentally by paternity studies (allozyme profiles, microsatellites, etc.), can be broken down into various elements (Marshall and Folsom, 1991):

Pre-pollen mechanisms: Prevention of self-fertilization to a certain extent and optimization of the departure (and reception) of pollen, as well as the development of strategies such as the development of pollinia (Asclepiadaceae, Orchidaceae), which make self-fertilization difficult.

Post-pollen mechanisms

— Pre-zygotic mechanisms involve control of pollen grain germination (receptivity of the stigma), competition between pollen grains during growth of the pollen tube, and the fertilization of one ovule, which is not the first but very often another one (sometimes none, as in radish).

— Post-zygotic mechanisms include abortion of seeds or fruits. Although the triggering of abortion of embryos is not clear, it is not a rare process. In the Rosaceae, the arrest of development of an embryo and its death originate from a deposit of callose and lignin in the transfer zone, which bocks the mother-daughter exchanges. In many species, embryos resulting from an interspecific cross abort, following processes that are still to be understood.[1]

Incompatibilities: gametophytic and sporophytic.

The choice of mate has an impact on the fitness of the progeny, because the embryo's resistance to abortion could be an inherited trait. That would mean that the control of sexual reproduction as we have discussed before increases the reproductive success of descendants. In greenhouse experiments with radish, for example, pollen grain donors that are less sensitive to seed abortion produced individuals having a significantly higher mass than the average (Marshall and Folsom, 1991). That is, the descendants grow better and probably have a higher survival rate (and thus fitness). Nevertheless, this relationship between choice and fitness is still to be clarified.

At the global scale, we observe a great diversity of Angiosperm species, which originates from a variety of factors (many of them unknown): we have mentioned the strategies of dispersal and dissemination but also underlined the importance of hybridization, or other modes of speciation. The adaptive radiation of Angiosperms led them to dominate the terrestrial plant world. Has the architecture of the Angiosperm flower been a pivot of this radiation? Probably yes: the sealing of ovular locules improved the protection of the embryo, and the sophistication of the gynoecium with a style, stigma, and syncarpy improved the control of fertilization and the production of complex fruits. Moreover, the androecium evolved towards an optimization and even specialization of pollination. The attractive function of sterile whorls improved greatly. Moreover, the development of self-incompatibility, as well as floral sexualization, considerably promoted cross-fertilization. All these anatomical and physiological factors have probably contributed to increasing the probability of cross-fertilization and the level of genetic diversity of Angiosperm populations (Heslop-Harrison, 1975), although self-fertilization does not automatically induce a low genetic diversity (see Chapter 5). Curiously, while self-incompatibility occurs frequently in plants, fertilization in animals is controlled more by means of dioecism (gonochorism). Other floral elements, which we have not addressed, have also contributed to the evolutionary success of Angiosperms, such as the shape and number of apertures in pollen grains.

---

[1]In *Datura*, it seems clear that the embryo is killed by the production of a substance toxic to it.

Finally, we note that, despite the multiplicity of aspects relating to floral biology, the evolution of Angiosperms is in general strikingly directed towards sexual reproduction—i.e., towards fertilization and dissemination—which is marked (1) by the existence of co-evolutions and (2) as we have already mentioned, by an orientation towards cross-fertilization. Consequently, as has been emphasized by Dilcher (2000), the characters of flowers, or even of fruits and seeds, clearly subjected to selection are probably the best criteria to classify this group.

The field of floral biology is far from being conquered. There are numerous uncertainties, notably about the genetic determination of floral development. The identification of genes and their modes of expression will undoubtedly, in the years to come, allow us to understand the establishment of the floral organization plan. The use of plant models other than *Arabidopsis, Antirrhinum*, or petunia must in this context be a source of information about the evolution of floral architecture and the development of the diversity of Angiosperm species.

Bibliography

Aizen, Basilio. Sex differential nectar secretion in protandrous *Alstromeria aurea* (Alstromeriaceae): is production altered by pollen removal and receipt? *Am. J. Bot.,* 85, 245-250, 1998.

Alvarez, Smyth. *Crabs claw* and *spatula*, two *Arabidopsis* genes that control carpel development in parallel with agamous. *Development*, 126, 2377-2386, 1999.

Ambrose, Lerner, Ciceri, Padilla, Yanofsky, Schmidt. Molecular and genetic analyses of the *Silky1* gene reveal conservation in floral organ specification between Eudicots and Monocots. *Mol. Cell,* 5, 569-579, 2000.

Angenent, Colombo. Molecular control of ovule development. *Trends Plant Sci.*, 1(7), 228-232, 1996.

Anstett, Hossaert-McKey, Kjellberg. Figs and fig pollinators: evolutionary conflicts in a coevolved mutualism. *TREE*, 12(3), 94-99, 1997.

Arroyo, Barrett. Discovery of distyly in narcissus (Amaryllidaceae). *Am. J. Bot.*, 87, 748-751, 2000.

Ayase, Schiestl, Paulus, Lofstedt, Hansson, Ibarra, Francke. Evolution of reproductive strategies in the sexually deceptive orchid *Ophrys sphegodes*: how does flower-specific variation of odor signals influence reproductive success? *Int. J. Org. Evol.*, 54(6), 1995-2006, 2000.

Barkman, Chenery, McNeal, Lyons-Weiler, Ellisens, Moore, Wolfe, de Pamphilis. Independent and combined analyses of sequences from all the three genomic compartments converge on the root of flowering plant phylogeny. *PNAS*, 97, 13166-13171, 2000.

Barret, Charlesworth. Effect of a change in the level of inbreeding on the genetic load. *Nature*, 352, 522-524, 1991.

Barret, Eckert. Variation and evolution of mating system in seed plants. In *Biological Approaches and Evolutionary Trends in Plants* (S. Kawano, ed.). Academic Press, pp. 229-254, 1990.

Barrrett. The evolution of mating strategies in flowering plants. *Trends Plant Sci.*, 3(9), 335-340, 1998.

Barrett, Harder. Ecology and evolution of plant mating. *TREE*, 11(2) 73-78, 1996.

Baum, Eshed, Bowman. The *Arabidopsis* nectary is an ABC-independent floral structure. *Development*, 128, 4657-4667, 2001.

Bawa. Plant-pollinator interactions in tropical rain forests. *Annu. Rev. Ecol. Syst.*, 21, 399-422, 1990.

Bawa. Pollination, seed dispersal, and diversification of Angiosperms. *TREE*, 10(8), 311-312, 1995.

Becker, Kaufmann, Freialdenhoven, Vincent, Li, Saedler, Theissen. A novel MADS-box gene subfamily with a sister group relationship to class B floral homeotic genes. *Mol. Genet. Genomics*, 266, 942-950, 2002.

Bertin. Theoretical consequences of heterogeneous pollen transport by animals. *Ecology*, 78(3), 962-963, 1997.

Boose. Sources of variation in floral nectar production in *Epilobium canum* (Onagraceae): implications for natural selection. *Oecologia*, 110(4), 493-500, 1997.

Bowe, Coat, de Pamphilis. Phylogeny of seed plants based on all three genomic compartments: extant gymnosperms are monophyletic and Gnetales' closest relatives are conifers. *PNAS*, 97(8), 4092-4097, 2000.

Bremer. Early cretaceous lineages of monocot flowering plants. *PNAS*, 97(9), 4707-4711.

Briggs, Walters. *Plant Variation and Evolution.* Cambridge University Press, pp. 152-283, 1997.

Broyles, Wyatt. Paternity analysis in a natural population of *Asclepias exaltata*: multiple paternity, functional gender, and the "pollen donation" hypothesis. *Evolution*, 44, 1454-1468, 1990.

Broyles, Wyatt. A re-examination of the pollen-donation hypothesis in an experimental population of *Asclepias exaltata. Evolution*, 49(1), 89-99, 1995

Brunet. Male reproductive success and variation in fruit and seed set in *Aquilegia caerulea* (Ranunculaceae). *Ecology*, 77(8), 2458-2471, 1996.

Brunet, Charlesworth. Floral sex allocation in sequentially blooming plants. *Evolution*, 49(1), 70-79, 1995.

Bui, O'Neill. Three ACC synthase genes regulated by primary and secondary pollination signals in orchid flowers. *Plant Physiol.*, 116, 419-428, 1998.

Burd, Allen. Sexual strategies in wind-pollinated plants. *Evolution*, 42, 403-407, 1988.

Campbell, Waser, Price. Mechanisms of hummingbird-mediated selection for flower width in *Ipomopsis aggregata. Ecology*, 77(5), 1463-1472, 1996.

Campillo, Bennett. Pedicel breakstrength and cellulase gene expression during tomato flower abscission. *Plant Physiol.*, 111, 813-820, 1996.

Carroll, Pallardy, Galen. Drought stress, plant water status, and floral trait expression in fireweed, *Epilobium angustifolium* (Onagraceae). *Am. J. Bot.*, 99, 438-446, 2001.

Charlesworth. The effect of investment in attractive structure on allocation to male and female functions in plants. *Evolution*, 41, 948-968, 1987.

Charnov. *The Theory of Sex Allocation.* Princeton University Press, 1982.

Charnov. See Freeman, Harper and Charnov, 1982.

Chaw, Parkinson, Cheng, Vincent, Palmer. Seed plant phylogeny inferred from all three plant genomes: monophyly of extant gymnosperms and origin of Gnetales from conifers. *PNAS*. 97(8), 4086-4091, 2000.

Chen, Meyerowitz. *Hua1* and *Hua2* are two members of the floral *agamous* pathway. *Mol. Cell*, 3, 349-360, 1999.

Clegg, Durbin. Flower color variation: a model for the experimental study of evolution. *PNAS*, 97(13), 7016-7023, 2000.

Crepet. Progress in understanding angiosperm history, success, and relationships: Darwin's abominably "perplexing phenomenon". *PNAS*, 98(24), 12939-12941, 2000.

Cronk, Moller. Genetics of floral symmetry revealed. *TREE*, 12(3), 85-86, 1997.

Davis, Pylatuik, Paradis, Low. Nectar carbohydrate production and composition vary in relation to nectary anatomy and location. *Planta*, 205(2), 305-318, 1998.

Devlin, Kay. Flower arranging in *Arabidopsis. Science*, 288, 1600-1602, 2000.

Deyholos, Sieburth. Separable whorl-specific expression and negative regulation by enhancer elements within the AGAMOUS second intron. *The Plant Cell*, 12, 1799-1810, 2000.

Diggle. Architectural effects and the interpretation of patterns of fruit and seed development. *Annu. Rev. Ecol. Syst.*, 26, 531-552, 1995.

Dilcher. Toward a new synthesis: major evolutionary trends in the angiosperm fossil record. *PNAS*, 97(13), 7030-7036, 2000.

Dixit, Rizzo, Nasrallah, Nasrallah. The Brassica MIP-MOD gene encodes a functional water channel that is expressed in the stigma epidermis. *Plant Mol. Biol.*, 45, 51-62, 2001.

Donoghue, Ree, Baum. Phylogeny and the evolution of flower symmetry. *Trends Plant Sci.*, 3(8), 311-317, 1998.

Dudareva, Murfitt, Mann, Gorenstein, Kolosova, Kish, Bonham, Wood. Developmental regulation of methyl benzoate biosynthesis and emission in snapdragon flowers. *The Plant Cell*, 12, 949-961, 2000.

Dudareva, Pichersky. Biochemical and molecular genetic aspects of floral scents. *Plant Physiol.*, 122, 627-634, 2000.

Eckert, Carter. Flowers produce variations in color saturation by arranging petals at oblique and varying angles. *J. Opt. Soc. Am. Opt. Im. Sci. Vis.*, 17(5), 825-830, 2000.

Ed Echeverria. Vesicle mediated solute transport between the vacuole and the plasmalemma. *Plant Physiol.*, 123, 1217-1225, 2000.

Emms, Stratton, Snow. The effect of inflorescence size on male fitness: experimental tests in the andromonoecious lily, *Zygadenus paniculatus. Evolution*, 51(5), 1481-1489, 1997.

Endress. *Diversity and Evolutionary Biology of Tropical Flowers.* Cambridge University Press, 1994.

Fahn. Ultrastructure of nectaries in relation to nectar secretion. *Am. J. Bot.*, 66, 977-985, 1979.

Ferdy, Despres, Godelle. Evolution of mutualism between globeflowers and their pollinating flies. *J. Theor. Biol.*, 217(2), 219-234, 2002.

Fishbein, Venable. Evolution of inflorescence design: theory and data. *Evolution*, 50(6), 2165-2177, 1996.

Fisher. Average excess and average effect of a gene substitution. *Ann. Eugen.*, 11, 53-63, 1941.

Freeman, Harper, Charnov. Sex change in plants: old and new observations and new hypothesis. *Oecologia*, 47, 212-232, 1980.

Freudenstein, Rasmussen. What does morphology tell us about orchid relationships? A cladistic analysis. *Amer. J. Bot.*, 86, 225-248, 1999.

Gandolfo, Nixon, Crepet, Stevenson, Friis. Oldest known fossils of monocotyledons. *Nature*, 394, 118-119, 1998.

Gocal, King, Blundell, Schwartz, Andersen, Weigel. Evolution of floral meristem identity genes: Analysis of *Lolium temulentum* genes related to *apetala-1* and *leafy* of *Arabidopsis. Plant Physiol.*, 125, 1788-1801, 2001

Gonzales-Bosch, del Campillo, Bennett. Immunodetection and characterization of tomato endo-beta-1,4-glucanase Cel 1 protein in flower abscission zones. *Plant Physiol.*, 114, 1541-1546, 1997.

Griffin, Mavraganis, Eckert. Experimental analysis of protogyny in *Aquilegia canadensis* (Ranunculaceae). *Am. J. Bot.*, 87, 1246-1256, 2000.

Guttierrez-Cortines, Davies. Beyond the ABCs: ternary complex formation in the control of floral organ identity. *Trends Plant Sci.* 5(11), 471-476, 2000.

Haig, Westoby. On limits to seed production. *Am. Nat.*, 131, 757-759, 1988.

Halle, Oldeman. *Essai sur l'Architecture et la Dynamique des Arbres Tropicaux.* Masson, Paris, 1970.

Harder, Wilson. Theoretical consequences of heterogeneous transport conditions for pollen dispersal by animals. *Ecology*, 79(8), 2789-2807, 1998.

Hasebe. Evolution of reproductive organs in land plants. *J. Plant Res.*, 112, 463-474, 1999.

Hempel, Welch, Feldman. Floral induction and determination: where is flowering control? *Trends Plant Sci.*, 5(1)17-21, 2000.

Herrera. The role of colored accessory bracts in the reproductive biology of *Lavandula stoechas. Ecology*, 78(2), 494-504, 1997.

Heslop-Harrison. Incompatibility and the pollen-stigma interaction. *Ann. Rev. Plant. Physiol.*, 26, 403-425, 1975.

Holsinger. Mass-action models of plant mating systems: the evolutionary stability of mixed mating systems. *Am. Nat.*, 138, 606-622, 1991.

Horridge. Bees see red. *TREE*, 13(3), 87-88, 1998.

Howell. *Molecular Genetics of Plants.* Cambridge University Press, 1998.

Hudson. Plant symmetry. *Annu. Rev. Ecol. Syst.*, 30, 364-365, 1999.

Ikeda, Nasrallah, Dixit, Preiss, Nasrallah. An aquaporin-like gene required for the *Brassica* self-incompatibility response. *Science*, 276, 1564-1566, 1997.

Jack. New members of the floral organ identity agamous pathway. *Trends Plant Sci.*, 7(7), 286-287, 2002.

Johnson, Nilsson. Pollen carryover, geitonogamy, and the evolution of deceptive pollination systems in orchids. *Ecology*, 80(8), 2607-2619, 1999.

Jones. *Introduction to Floral Mechanism.* Blackie & Son, London, 1950.

Jones, Reithel. Pollinator-mediated selection on a flower color polymorphism in experimental populations of *Antirrhinum* (Scrophulariaceae). *Am. J. Bot.*, 88, 447-454, 2001.

Jones, Woodson. Differential expression of the three members of the ACC synthase gene family in carnation. *Plant Physiol.*, 119, 755-764, 1999.

Jordan, Franklin, Franklin-Tong. Evidence for DNA fragmentation triggered in the self-incompatibility response in pollen of *Papaver rhoeas. The Plant J.*, 23(4), 471-479, 2000.

Judd, Campbell, Kellogg, Stevens. *Plant Systematics: a Phylogenetic Approach.* Sinauer Publishers, 1999.

Kahana, Silberstein, Kessler, Goldstein, Perl-Treves. Expression of ACC oxidase genes differs among sex genotypes and sex phases in cucumber. *Plant. Mol. Biol.*, 41, 517-528, 1999.

Kater, Colombo, Franken, Busscher, Masiero, van Lookeren-Campagne, Angenent. Multiple AGAMOUS homologs from cucumber and petunia differ in their ability to induce reproductive organ fate. *The Plant Cell*, 10, 171-182, 1998.

Kater, Franken, Carney, Colombo, Angenent. Sex determination in the monoecious species cucumber is confined to specific floral whorls. *The Plant Cell.*, 13, 481-493, 2001.

Kjellbom, Larsson, Johansson, Karlsson, Johanson. Aquaporins and water homeostasis in plants. *Trends Plant Sci.*, 4(8), 308-314, 1999.

Klinkhamer, de Jong, Metz. Sex and size in cosexual plants. *TREE* 12(7), 260-265, 1997.

Koopowitz, Marchant. Postpollination nectar reabsorption in the African epiphyte *Aerangis verdickii* (Orchidaceae). *Am. J. Bot.*, 85, 508-512, 1998.

Koriba. On the periodicity of tree growth in the tropics with reference to the mode of branching, the leaf fall, and the formation of the resting bud. *Garden's Bull. Singapore*, 17(1), 11-81, 1958.

Kramer, Irish. Evolution of the petal and stamen developmental programs: evidence from comparative studies of the lower eudicots and basal Angiosperms. *Int. J. Plant Sci.*, 161(Suppl.), S29-S40, 2000.

Krizek, Prost, Macias. AINTEGUMENTA promotes petal identity and acts as a negative regulator of AGAMOUS. *The Plant Cell*, 12, 1357-1366, 2000.

Law, Bronstein, Ferriere. On mutualists and exploiters: plant-insect coevolution in pollinating seed-parasite systems. *J. Theor. Biol.*, 212(3), 373-389, 2001.

Levin, Brack. Natural selection against white petals in phlox. *Ecology*, 49(5), 1017, 1995.

Lippock, Gardine, Williamson, Renner. Pollination by flies, bees, and beetles of *Nuphar ozarkana* and *N. advena* (Nympheaceae). *Am. J. Bot.*, 87, 898-902.

Liu, Franks, Klink. Regulation of gynoecium marginal tissue formation by *leunig* and *aintegumenta. The Plant Cell*, 12, 1879-1892, 2000.

Lloyd. Some reproductive factors affecting the selection of self-fertilization in plants. *Am. Nat.*, 113, 67-79, 1979.

Lloyd. Modification of gender in seed plants in varying condition. *Evol. Biol.*, 17, 255-388, 1984.

Lloyd. A general principle for the allocation of limited resources. *Evol. Ecol.*, 2, 175-187, 1988.

Lloyd. *Floral Biology: Studies on Floral Evolution in Animal Pollinated Plants*. Chapman & Hall, New York, 1996.

Lloyd, Wells. Reproductive biology of a primitive angiosperm, *Pseudowintera colorata* (Winteraceae) and the evolution of pollination systems in the Anthophyta. *Plant Syst. Evol.*, 181, 77-95, 1992.

Luttge, Higinbotham. *Transport in Plants.* Springer-Verlag, 1969.

Lyons-Weiler, Hoelzer, Tausch. Relative apparent synapomorphy analysis (RASA): the statistical measurement of phylogenetic signal. *Mol. Biol. Evol.*, 13(6), 749-757, 1996.
Lyons-Weiler, Hoelzer, Tausch. Optimal outgroup analysis. *Biol. J. Lin. Soc.*, 64, 493-511, 1998.
Marshall, Folsom. Mate choice in plants. *Ann. Rev. Ecol. Syst.*, 22, 37-63, 1991.
Maynard Smith. *Evolution and the Theory of Games.* Cambridge University Press, 1982.
McConnell, Barton. Leaf polarity and meristem formation in *Arabidopsis. Development*, 125, 2935-2942, 1998.
Melendez-Ackerman, Campbell, Waser. Hummingbird behaviour and mechanisms of selection on flower color in *Ipomopsis. Ecology*, 78(8), 2532-2541, 1997.
Meyerowitz. La genetique des fleurs. *Pour la Science*, 207, 58-66, 1995.
Moller. Bumblebee preference for symmetrical flowers. *PNAS*, 92, 2288-2292, 1995.
Morgan, Schoen. The role of theory in an emerging new plant reproductive biology. *TREE*, 12(6), 231-234, 1997.
Morris. Mutualism denied? Nectar robbing bumble bees do not reduce female or male success of bluebells. *Ecology*, 77(5), 1451-1462, 1996.
Mulcahy. The rise of the Angiosperms: a genecological factor. *Science*, 206, 20-23, 1979.
Murcia, Feinsinger. Interspecific pollen loss by hummingbirds visiting flower mixtures: effects of floral architecture. *Ecology*, 77(2), 550-560, 1996.
Mutikainen, Delph. Inbreeding depression in gynodioecious *Lobelia siphilitica*: among-family differences override between-morph differences. *Evolution*, 52(6), 1572-1582, 1998.
Nasrallah. Cell-cell signalling in the self-incompatibility response. *Curr. Op. Plant Biol.*, 3, 368-373, 2000.
Nasrallah, Nishio, Nasrallah. The self-incompatibility genes of *Brassica*: expression and use in genetic ablation of floral tissues. *Annu. Rev. Plant Physiol. Plant Mol. Biol.*, 42, 393-422, 1991.
Neal, Dafni, Giurfa. Floral symmetry and its role in plant-pollinator systems. *Annu. Rev. Ecol. Syst.*, 29, 345-373, 1998.
Olmstead, de Pamphilis, Wolfe, Young, Elissons, Reeves. Disintegration of Scrophulariaceae. *Am. J. Bot.*, 88(2), 348-361, 2001.
O'Neill. Pollination regulation of flower development. *Annu. Rev. Plant Physiol. Plant Mol. Biol.*, 48, 547-574, 1997.
Owens, Takaso, Runions. MADS box genes in coniferous plants. *Trends Plant Sci.*, 3, 459-460, 1998.
Palanivelu, Preuss. Pollen tube and axon guidance: parallels in tip growth mechanisms. *Trends in Cell Biol.*, 10, 517-523, 2000.
Parcy, Nilsson, Busch, Lee, Weige. A genetic framework for floral patterning. *Nature*, 39, 561-566, 1998.
Pelaz, Ditta, Baumann, Wisman, Yanofsky. B and C organ identity functions require *sepallata* MADS-box genes. *Nature*, 405, 200-203, 2000.
Pellmyr, Leebens-Mack. Forty million years of mutualism: evidence for Eocene origin of yucca-yucca moth association. *PNAS*, 96, 9178-9183, 1999.
Powell. Interrelationships of yuccas and yucca moths. *TREE*, 7(1), 10-14, 1992.
Purugganan, Boyles, Suddith. Variation and selection at the cauliflowers floral homeotic gene accompanying the evolution of domesticated *Brassica oleracea. Genetics*, 155, 855-862.
Raven, Weyers. Significance of epidermal fusion and intercalary growth for angiosperm evolution. *Trends Plant Sci.*, 6(3), 111-113, 2001.
Regal. Ecology and evolution of flowering plant dominance. *Science*, 196, 622-629, 1977.
Richter, Schranner. Leaf arrangement. Geometry, morphogenesis and classification. *Naturwissenschaften*, 65, 319-327, 1978.
Rickels, Renner. Species richness within families of flowering plants. *Evolution*, 48(5) 1619-1636, 1994.

Rigola, Mizzi, Ciampolini, Sari-Gorla. CaMADS1, an AGAMOUS homologue from hazelnut, produces floral homeotic conversion when expressed in *Arabidopsis*. *Sex. Plant Reprod.*, 13(4), 185-191, 2001.

Roy, Widmer. See Jones.

Rubinstein. Regulation of cell death in flower petals. *Plant Mol. Biol.*, 44: 303-318, 2000.

Sakai. Role of SUPERMAN in maintaining *Arabidopsis* floral whorl boundaries. *Nature*, 378, 199-203, 1995.

Sakai, Harada. Does the trade-off between growth and reproduction select for female-biased sexual allocation in cosexual plants? *Evolution*, 52(4), 1204-1207, 1998.

Schemske, Bradshaw. Pollinator preference and the evolution of floral traits in monkeyflower *(Mimulus)*. *PNAS*, 96(32), 11910-11915, 1999.

Schill, Baumm, Wolter. Vergleichende Mikromorphologie der Narbenoberflachen bei den Angiospermen; Zusammenhange mit Pollenoberflachen bei heterostylen Sippen. *Plant Syst. Evol.*, 148, 185-214, 1985.

Schneitz, Balasubramanian, Scheifthaler. Organogenesis in plants: the molecular and genetic control of ovule development. *Trends Plant Sci.*, 3(12), 468-472, 1998.

Schoen, Dubuc. The evolution of inflorescence size and number: a gamete packaging strategy in plants. *Am. Nat.*, 135, 841-857, 1990.

Seger. Evolution of sexual systems and sex allocation in plant when growth and reproduction overlap. *Proc. R. Soc. London Ser. B* 263, 833-841, 1996.

Sheppard, Brunner, Krutovskii, Rottman, Skinner, Vollmer, Strauss. A DEFICIENS homolog from the dioecious tree black cottonwood is expressed in female and male floral meristems of the two-whorled, unisexual flowers. *Plant Physiol.*, 124, 627-640, 2000.

Shoen, Ashman. The evolution of floral longevity: resource allocation to maintenance versus construction of repeated parts in modular organisms. *Evolution*, 49(1), 131-139, 1995.

Smithson, Macnair. Negative frequency-dependent selection by pollinators on artificial flowers without rewards. *Evolution*, 51(3), 715-723, 1997.

Soltis, Soltis, Chase. Angiosperm phylogeny inferred from multiple genes as a tool for comparative biology. *Nature*, 402, 402-404, 1999.

Soltis, Soltis, Mort, Chase, Sarolainen, Hoot, Morton. Inferring complex phylogenies using parsimony: an empirical approach using three large DNA data sets for Angiosperms. *Syst. Biol.*, 47, 32-42, 1998.

Southerton, Marshall, Mouradov, Teasdale. Eucalypt MADS-box genes expressed in developing flowers. *Plant Physiol.*, 118, 365-372, 1998.

Spaethe, Tautz, Chittka. Visual constraints in foraging bumblebees: flower size and color affect search time and flight behavior. *PNAS*, 98(7), 3898-3903, 2001.

Sun, Dilcher, Zheng, Zhou. In search of the first flower: a Jurassic angiosperm, *Archaefructus*, from Northeast China. *Science*, 282, 1692-1695, 198.

Sung, Yu, An. Characterization of the MdMADS2, a member of the SQUAMOSA subfamily of genes in apple. *Plant Physiol.*, 120, 969-978, 1999.

Taiz and Zeiger. 2002. Plant Physiology, 3rd Ed. Sinauer Associates Inc, 2002.

Takasaki, Hatakeyama, Suzuki, Watanabe, Isogai, Hinata. SRK is the determinant of self-incompatibility specificity in the stigma. *Nature*, 403, 913-916, 2000.

Tang, Woodson. Temporal and spatial expression of ACC oxidase mRNA following pollination of immature and mature petunia flowers. *Plant Physiol.*, 112, 503-511, 1996.

Theissen, Becker, Di Rosa Kanno, Kim, Munster, Winter, Saedler. A short history of MADS-box genes. *Plant Mol. Biol.*, 42, 115-149, 2000.

Tiffney, Mazer. Angiosperm growth habit, dispersal and diversification reconsidered. *Evol. Ecol.*, 9, 93-117, 1995.

Till-Bottraud, Venable, Dajoz, Gouyon. Selection on pollen morphology: a game theory model. *Am. Nat.*, 144(3), 395-411, 1994.

Vallade. *Structure et Développement de la Plante*. Dunod Editions, 1999.

Verbelen, Tao. Mobile arrays of vacuole ripples are common in plant cells. *Plant Cell Rep.*, 17, 917-920, 1998.

Vishnevetsky, Ovadis, Vainstein. Carotenoid sequestration in plants: the role of carotenoid-associated proteins. *Trends Plant Sci.*, 4(6), 232-235, 1999.

Vogel. Die Oelblumensymbiosen—Parallelismus und andere Aspeckte ihrer Entwicklung in Raum und Zeit. *Z. Zool. Syst. Evol. Forsch.*, 26, 341-362, 1988.

Waser, Chittka. Bedazzled by flowers. *Nature*, 394, 835-836, 1998.

Weigel, Clark. Sizing up the floral meristem. *Plant Physiol.*, 112, 5-10, 1996.

Wiehler. A synopsis of the neotropical Gesneriaceae. *Selbyana*, 6, 1-219, 1983.

Wolfe, Krstolic. Floral symmetry and its influence on variance in flower size. *Am. Nat.*, 154, 484-487, 1999.

Woltering, Somhorst, van der Veer. The role of ethylene in interorgan signalling during flower senescence. *Plant Physiol.*, 109, 1219-1225, 1995.

Xu, Hanson. Programmed cell death during pollination induced petal senescence in petunia. *Plant Physiol.*, 122, 1323-1334, 2000.

Yalovsky, Rodriguez-Concepcion, Bracha, Toledo-Ortiz, Gruissem. Prenylation of the floral transcription factor APETALA 1 modulates its function. *The Plant Cell.* 12, 1257-1266, 2000.

Yu, Goh. Identification and characterization of three orchid MADS-box genes of the AP-1/AGL-9 subfamily during floral transition. *Plant Physiol.*, 123, 1325-136, 2000.

Zhao, Yu, Chen, Ma. The *ask1* gene regulates B function gene expression in cooperation with *UFO* and *leafy* in *Arabidopsis. Development*, 128, 2735-2476, 2001.

General Index

A

B

C

D

E

F

G

H

I

T

U

W

Z

Botanical Index

H

I

J

L

M

N

O

P

R

S

T

Y

Z